Concentration

Concentration

Staying Focused in Times of Distraction

Stefan Van der Stigchel

Translated by Danny Guinan

The MIT Press

Cambridge, Massachusetts | London, England

This publication has been made possible with financial support from the Dutch Foundation for Literature.

N **ederlands**
letterenfonds
dutch foundation
for literature

This book was set in ITC Stone Serif Std and ITC Stone Sans Std by Toppan Best-set Premedia Limited.

Library of Congress Cataloging-in-Publication Data

Names: Stigchel, Stefan van der, author.
Title: Concentration : staying focused in times of distraction / Stefan Van der Stigchel ; translated by Danny Guinan.
Other titles: Concentratie. English
Description: Cambridge, Massachusetts : The MIT Press, [2020] | Includes bibliographical references and index.
Identifiers: LCCN 2019025808 | ISBN 9780262538565 (paperback) | ISBN 9780262357838 (ebook)
Subjects: LCSH: Attention. | Cognitive psychology. | Social media--Psychological aspects.
Classification: LCC BF322 .S7513 2020 | DDC 153.7/3--dc23
LC record available at https://lccn.loc.gov/2019025808

10 9 8 7 6 5 4 3 2

My experience is what I agree to attend to. Only those items which I notice shape my mind.

—William James, *The Principles of Psychology*, vol. I

Contents

Prologue

In the first century AD, the Roman philosopher Lucius Annaeus Seneca came to the conclusion that there was too much information out there. There were so many books in circulation at the time that Seneca was worried that the abundance of information was becoming a major source of distraction. The same problem resurfaced in Germany in the 18th century when there was serious talk of a *Bücherseuche*—a plague of books. The fear was that because there was so much information available, it might have a damaging effect on people's concentration.

Each new kind of media brings with it new questions, from the invention of the printing press to the rise of today's digital society. Concerns as to the effect of increasing amounts of information on our concentration are nothing new, and each time a new advance is made, people fear that we may be reaching the limits of our brain's capacity. Up to now this has not been the case, but the fear that we are fast approaching that moment is more understandable now than ever before. Our society is currently undergoing massive changes. Whereas in the eighties and nineties we spoke about information in terms of kilobytes and megabytes, today it's all zettabytes and yottabytes. The pile of information is growing ever higher and it is encroaching more and more upon our daily lives thanks to the speedier networks, advanced storage methods, and faster processors that we carry around in our pockets. Google and Facebook know exactly what your interests are because they can track your online behavior, which means that the advertisements you see are selected on the basis of your personal tastes. (I always give myself a bit of a fright every time I accidentally click on a website banner because I know that Google will react by sending me ads based specifically on that click.)

The incredibly fast pace of digital change and the increasing amount of available information means that right now we find ourselves in the middle of an attention crisis. If you don't believe me, just take a look at the number of books and articles that are published every week warning us of the disastrous consequences of social media and smartphones for our creativity, productivity, and concentration. The market for books on digital detox is huge because of the stress people are experiencing as a result of the continuous distraction of social media. At the same time, the number of traffic accidents continues to rise, thanks to our addiction to the smartphone, and the number of children who have trouble sleeping because they are glued to their screens at night is on the rise, too. Ever since the publication of my previous book, *How Attention Works*, in 2019, in which I explained the basic principles of attention, I have been asked the same question over and over again during interviews and readings: what is the effect of a rapidly changing society on our ability to concentrate? This appears to be a very important question for many. A lot of people find it difficult to concentrate these days because of the consistently high level of distraction and the never-ending stream of information that washes over us every minute of every day.

Today's children are growing up in a world that is a lot different from how the world was only twenty years ago. Teenagers are massive consumers of social media and other attention distractors, and this continuous yet subtle pulling on a teenager's attention happens in a phase when their brain has not yet fully developed, with the result that they end up using their attention the wrong way. All it takes is one moment of mental weakness or boredom and out comes the smartphone. Just when they have settled down to concentrate on their studies, the sound of an incoming message is there to distract their attention again. It is no wonder that more and more teachers are complaining about how difficult it is becoming to attract and hold their students' attention. The competition from smartphones is simply overwhelming, even when they are banned from the classroom.

The smartphone is also becoming more and more intrusive in our homes. Even parents are finding it difficult to spend enough time with their children. One bleep from their phone can cause them to abandon the Lego house they were building with their child, and it can then take an age before they are able to refocus their attention on what they were doing. Sometimes even that is too much to ask. The bleep might be a message from a colleague that either spoils the mood or forces Mum or Dad to turn their attention to their work again.

Not all of the doom scenarios sketched in self-help books and the media are accurate, of course. A lot of nonsense is being spouted these days, and there is absolutely no reason to believe that we are all going to end up suffering from digital dementia or that our intelligence levels are eventually destined to plummet. However, this does not mean that everything is hunky-dory either. Our tendency to multitask and the endless stream of information we are subjected to means that many of us have great difficulty concentrating. But there is also some good news: the rise of the digital society has come paired with another development, namely our increased knowledge of our behavior and of the brain. Our understanding of how concentration works is getting better and better, and this is helping us to figure out how to use it in the most optimal manner possible. The solution to the problem of what our society needs to do to be able to deal with the never-ending stream of information lies in our own hands. It is simply a matter of making the right choices.

That said, the choices we must make to prevent ourselves from spiraling into a full-blown attention crisis are not easy ones. In this book, I will draw a clear distinction between the sender and the receiver of information. Both will have to make adjustments if we want to preserve our ability to concentrate. Every day we all play the role of both sender and receiver. If you have to give a presentation or make a clip for YouTube, you are a sender, and you are a receiver when you attend a lecture or read a book. On the one hand, the sender has to work very hard to retain the attention of the receiver by offering them information in a way that makes it easy for them to concentrate. On the other hand, the receiver has to use his or her attention wisely and not allow themselves to become distracted. This requires not only suitably designed working and learning spaces, but also that the receiver does their best to train their concentration and keep their brain in good shape. Luckily, science is not short on solutions, and I will do my best to explain them to you in this book.

The Attention Crisis

You may have noticed how more and more companies have started using slogans such as "Attention makes everything more beautiful" (IKEA), "We like to pay more attention to our clients" (Eyewish opticians), and "We pay close attention to your future" (Aegon Insurance). In fact, it is remarkable just how many advertising campaigns these days are focused almost exclusively on the subject of attention. Businesses and advertisers are well aware of the important role attention plays in today's market and they spare no effort in attracting the eyes and ears of the consumer in order to get their information across.

Contrary to IKEA's claim, attention does not make everything more beautiful, but it does ensure that the brain selects certain data from the constant and never-ending stream of information. Attention works like a filter, and it is only the information that manages to get through this filter that ends

up being fully processed by the brain. All other information is simply ignored. Everything we know is determined,
therefore, by where we focus our attention. Our brain has
to react extremely quickly to the incredibly complex world
around us, and this means that we are constantly required
to make decisions. Who has the right of way at the intersection? How much attention should I pay to the message
blinking on my phone? The speed at which we must decide
often means that we have to make decisions without having the time to consider them fully. Good examples of this
are impulse buying or a spontaneous emotional outburst.
In these situations, our decisions are led by our initial reaction. We also draw on our knowledge of the world around us
when making these decisions, but that knowledge is based
on the things on which we have focused our attention in the
past. Attention therefore plays a crucial role in every single
decision that you make.

The Attention Economy

Attention is a valuable commodity and one we need to use
sparingly because we only have a limited amount of it at our
disposal at any given moment. Over the past few years, a
number of authors have published books on the rise of the
attention economy and its main item of trade: our very valuable attention. For instance, in his book *The Attention Merchants,* Tim Wu describes how the attention economy first
put down its roots in the newspaper world. When we read

a newspaper, we voluntarily offer our attention to the publisher of that paper, and the first to catch on to this was *The Sun* in New York. In 1833, newspapers were a luxury item aimed primarily at the elite and were sold at a price above the cost price. The owners of *The Sun* decided to take a different approach. They came up with the idea of selling the voluntary attention of its readers to the newspaper's advertisers. As a result, they were able to sell their paper for as little as one-sixth of the price charged by their competitors and generated most of their revenue by selling advertising space. The move was a resounding success and within one year *The Sun* had become the best-selling newspaper in New York.

It is every advertiser's dream: a large, fixed group of consumers with a clear profile spending a large portion of their attention on reading the newspaper each day. Using their knowledge of this target group, advertisers adjusted their messages to match the group's specific needs. This selling of voluntary attention eventually became the basis for today's media landscape. Ads are the most important source of income for companies like Facebook and Google, companies with which we interact every single day. The services they provide may appear free at first glance, but the price, in fact, is a significant chunk of our attention.

Advertising agencies know all the rules (and tricks) for attracting our attention and the advertisements they design to do so are becoming ever more refined. They tap into our basic emotions, the ones that cause us to regard naked people and scary situations, for example, as very "interesting." Eye-catching photos are used to grab our attention, and articles warning us about some calamity or other are almost always the most popular items on news sites. More than ever before, a fierce battle is raging for our attention—a battle that is often dirty and ruthless. Advertisers know exactly what they need to do to distract us. Want to watch a clip on YouTube? First you'll need to watch this thirty-second-long advertisement. Want to skip the ad? When you click on the "x" to close the ad, all it does is lead you to a pop-up window. And there goes another slice of your valuable attention.

Even the like button, which was originally introduced to add more positivity to social media, has become just another way of finding out more about users so that the algorithms for advertisements can be made even smarter. Likes also ensure that you are drawn back to social media because you will want to know whether your input is being appreciated or not, which in turn generates the kind of short-term reward that our brains are so fond of. Snapchat works by using a series of messages, or "streaks," that friends send to each other, and it requires you to send a message every day if you don't want to be the one responsible for breaking the chain. This can lead to enormous social pressure, something to which young people are particularly sensitive.

Attention in the Media

The attention economy determines to a large extent the way information is disseminated via the media. Throughout history, there have been major changes in the amount of attention we devote to the different kinds of media. This has had an enormous effect on our *attention rituals*. Around 1920 the radio was still more or less an ad-free zone. People believed there was no place on the radio for commercials, mostly because listening to the radio was seen as a family affair and a firm line should be drawn at the door of the living room when it came to exposure to advertising.

Thanks in part to the increasing popularity of radio, this attitude began to change and the radio, and later on the television, soon became the perfect medium for disseminating commercial messages. The fact that there were relatively few channels meant that certain broadcasters had enormous reach. Popular programs became known for their "attention spikes": people tuned in to certain shows in their millions and voluntarily surrendered all of their attention to a single broadcaster. Because there was so little choice and, just as significantly, the remote control had yet to be invented, this attention was very high in terms of its intensity. People were glued to the radio or TV set and devoted every ounce of their attention to the broadcast. This intense attention for one particular program disappeared when the remote control made its entrance. The new device ensured that there was less attention available for each individual show because people could and would switch between different channels without even knowing which shows were being broadcast. Today we watch television with our finger on the trigger, so to speak: one moment of weakness and our attention slips and we zap to the next channel in our perpetual search for new information.

When we switch between channels, we are allowing ourselves to be led by our impulses in deciding which channel we will watch. Instead of focusing our attention voluntarily on one particular stream of information, we jump, as if by reflex, from one channel to the next. Program makers have adapted their methods to match this behavior. TV shows from yesteryear now appear incredibly slow to us, but that is primarily because they didn't have the sword of Damocles that is the remote control hanging over their heads. This also applies to ads on Facebook's newsfeed, which usually begin to play automatically because they have a very small window in which to grab your attention.

The same phenomenon can be found in the world of music. With so much music available on Spotify nowadays, the first thirty seconds of a song are absolutely crucial. So much so that choruses tend to come earlier in many songs and intros often have no beats at all so that the listener can be enticed to stay tuned without becoming irritated. Otherwise, within a fraction of a second, they will have scrolled on down the page and clicked on a track that catches their eye and to which they may be prepared to devote their precious attention voluntarily for a longer period of time. It costs our brain more effort, after all, to focus on something for a relatively long period of time than to be led by whatever information comes in randomly from the outside world. The latter requires no control. In short, the attention economy ensures that information is disseminated at an increasingly faster pace.

Attention and Technology

The battle for our attention has become even fiercer as a result of technological developments and the huge increase in the flow of information. And the stimuli that come with this battle often distract us from the things that we really want to do. In the library where I am writing this book, I can see lots of students engrossed not in their studies but in Facebook, Twitter, and Instagram. Was that really what they had in mind when they were cycling to the library this morning? Shouldn't they be working on that all-important essay or reading one of their textbooks instead? Maybe they just became distracted by a notification on their smartphone alerting them to a new friend request, with the result that their attention is now firmly fixed on social media instead of on their essay. In that case it will take a lot of effort before they are able to drag themselves away their smartphone and focus on their work again.

Our cell phones have a lot to answer for in the current attention crisis. Between 1985 and 2010, the number of cell phone subscriptions in the United States rose from a modest 340,000 to a whopping 302,900,000. Even the function for which it was originally designed—making a phone call—has become a huge distraction in terms of the amount of attention we pay to the world around us. Although we are able to look around us when making a call, the attention it eats up means that we take in very little of what is actually going on in our immediate surroundings. This has been clearly demonstrated many times in controlled laboratory experiments, but a group of scientists from the Western Washington University decided to test the theory even further on the campus of their university in an attempt to reveal the extent to which the problem affects our daily lives. In their study, they observed 350 unsuspecting test subjects as they walked across the campus square and categorized them based on what they were doing: talking to another person, chatting on the phone, walking along on their own listening to music, or just walking in silence. Drawing their inspiration from the famous "invisible gorilla" experiment, the scientists introduced a clown cycling across the square on a unicycle to the scene.

Among the test subjects who were listening to music or walking in silence, one in three responded that they had seen a clown on a unicycle, and nearly 60 percent of those who were walking with a friend mentioned the clown. But among those who had been talking on their cell phone, only 8 percent spontaneously remembered the clown. Then the researchers asked a second question: "Did you see the clown on the unicycle?" With prompting, 71 percent of the people walking with a friend remembered the clown. The numbers were also higher for people listening to music (61 percent) and those who were walking alone (51 percent). But among those who had been talking on a cell phone, the ability to recall seeing the clown was very low, with only 25 percent of them saying that they had seen a clown on a unicycle. In addition, the subjects' walking behavior was also analyzed. The test subjects who were talking on their cell phone walked slower and changed direction more often. Many other studies have produced similar results. For example, one field study involving more than 500 test subjects demonstrated that pedestrians engrossed in their cell phones cross the road more slowly and look around them a lot less than pedestrians who are not busy on the phone.

These studies show that using a cell phone to make a call leads to reduced *situation awareness*. An extensive American study into the causes of accidents involving pedestrians concluded that the use of cell phones is becoming an increasingly important factor and that the number of accidents involving pedestrians talking on their cell phones had doubled in only a few years. These conclusions have been confirmed not only by the aforementioned observation studies but also by a number of experimental studies in which test subjects were asked to walk a particular route either with or without their cell phone. The test subjects talking on a cell phone took in less of the world around them and failed to notice many of the eye-catching objects along their route. It seems, therefore, that we are unable to concentrate fully both on our surroundings and the telephone conversation we happen to be conducting. Today, the omnipresence of the cell phone is partly responsible for the rise in traffic accidents, simply because we now pay less attention while in traffic.

Why Is the Cell Phone So Addictive?

We have seen just how much of an influence the cell phone can have on our lives. But why is it so addictive? This can best be explained by first giving a short introduction to the ways in which we acquire our behavior, based on a scientific movement known as *behaviorism*. Behaviorism was the first movement within the field of psychology to question the standard method of introspection in which test subjects themselves provided the information about what they felt and experienced during an experiment. For example, the introspection technique required test subjects to describe what they experienced while listening to a metronome. They would subsequently identify certain rhythms as being more pleasant than others. Behaviorists believed that these subjective findings were unreliable, so they started conducting experiments in which observable behavior could be measured. Behaviorism officially came into being in 1913 with the publication of the manifesto, *Psychology as the Behaviorist Views It*, by John Watson, in which the author suggested that psychological studies should focus their attention on "stimulus response relationships": people react to incoming information as if by reflex, like the way a light goes on when you flip the switch—a stimulus appears from the outside world and it leads automatically to a reaction, as if you were a machine. The subsequent relationships are formed through conditioning: the acquisition of reflex actions.

Conditioning was discovered by accident by Ivan Pavlov around 1890 when he was studying the production of saliva by dogs as a reaction to food. He noticed that the dogs began to salivate the moment he entered the room, even when he wasn't carrying any food with him. It appeared that the dogs had established an automatic link between his presence and feeding time. This had not been the case at first, but the more acquainted he and the dogs became, the stronger the link became, too. Eventually, they began to react to his presence in the same way they reacted to food. In the language of behaviorism, Pavlov himself had at first been a neutral stimulus before becoming a conditioned stimulus over time.

Pavlov's discovery was later researched more exhaustively using bells, among other objects, but the conclusion remained the same, regardless of whether it applied to dogs or humans: once we have started to associate a certain object with a reward, that association can be very hard to reverse. To do so we have to undergo a long period in which the object no longer leads to a reward. The same principle applies to punishment, for example. If you give a test subject an electric shock once or twice when they are viewing a certain color, the brain will literally get a fright each time the color is subsequently shown, even in the absence of the electric shock. According to behaviorists, we do not have any free will in this kind of situation but exhibit instead a reflex reaction to whatever we experience in our surroundings.

A reward can have a very addictive effect. When mice in a cage discover they will be given food if they press a button, they will keep on pressing that button, ad infinitum. A reward can also have an educational effect, and we use this very principle when raising our kids and teaching them in our schools. As long as certain behavior continues to be rewarded, we will continue to exhibit that behavior, even when it leads, momentarily, to no reward. When we instinctively pull out our cell phone to see if we have any new messages or open our mail on the computer we are, in fact, behaving like the conditioned mice above. In most cases, when we stop what we are doing so that we can check for new messages, what we are looking for is a reward. We experience each new message as a kind of reward, whether it is a quick hello from someone who is thinking of you, a funny video with a cat that someone wants to share with you, or even just a work-related email. It's all new information, so happy days!

We therefore associate the messages we receive on our cell phones and computers with the likelihood of receiving a reward. This association arises from the fact that, in the past, we have probably received messages that we experienced as pleasing (like the message from someone saying a quick hello) and interpreted them as a reward. Like the mice in the cage, we form an association between social media, for example, and a reward, such as a like or a message from a friend. And we continue to press the button—again, just like the mice—even when there is no real reason to do so. It has now become an addiction, and the association does not have to be confirmed each time anymore for it to be maintained. However, the fact that there is no guarantee that a reward will follow via email or social media is precisely why these rewards are so fickle. In the case of a variable reward, where the reward varies in terms of extent and frequency, it takes longer for the association to be learned, but also longer for it to be erased (a process known as *extinction*). If we always knew exactly when to expect a reward, we would only ever reach for our phones at those moments. Unfortunately, a rewarding message does not always arrive at a predictable moment.

The very same applies to emails. It is very easy—and free—to send an email, and so our mailboxes are often stuffed to overflowing, unlike back in the days when it required a lot more effort to send a message to someone (finding a sheet of paper, a pen, a postage stamp, a letterbox, etc.). All you need now is a thought and a moment to type out that thought, and off it goes. The mail is usually delivered only once a day, but an email can arrive at any time. This leads to the creation of an attention ritual whereby we get used to checking our email very frequently on the off chance that something new and interesting may have arrived. You can look forward to your mail arriving through the letterbox in the morning, but once it has been delivered, you know that you will not be receiving any more mail that day. Furthermore, you will not be inclined to write a reply immediately to the letters you have received and to run to the letterbox with them as quickly as your legs will carry you. Sending a letter takes time, and your correspondence partner won't expect you to reply to their letter immediately.

Sending an email couldn't be more different. We are all familiar with the kind of colleague who comes running over to your desk five minutes after they have sent you an email to ask whether you have read it yet. When you think about it, it really is quite amazing that someone can lay claim to a significant slice of your valuable attention simply by sending you an email or a text message.

Attention in Health Care

If there is one place where there is a growing realization that attention is a scarce and valuable commodity, it is the health care sector. An increasingly popular feature of health care protocols these days is what are known as *attention minutes*, a term that refers to the practice of reserving a certain amount of time to chat with a patient after they have been given the treatment they require. Home care workers are now often armed with a checklist indicating the number of minutes they may spend with their patients. For example: twenty minutes for tasks such as washing and dressing, and five minutes for a chat afterwards.

There is a debate currently ongoing within the health care sector about whether providing nonmedical attention to a patient forms a fundamental part of care provision or is something extra that can be added only when there is time left over to do so. According to some, health care can only remain affordable when care workers focus all of their attention on the medical care required. All other forms of attention required by the patient are then deemed to be the responsibility of the family or an extra service that can be provided when there is enough time left over. Given the fact that in the Netherlands, for example, increasingly less funds are being allocated to the care of the elderly, their attention needs can no longer be regarded as a purely public concern. Instead, it is the family that is expected to take responsibility for elder care. When they do so, care workers, who have little or no time for this task anyway, are more able to focus all of their attention on the essential health care they are trained to provide.

In scientific circles, however, a different voice is currently making itself heard, one that suggests there is more to good health care than just following the correct procedures for whatever professional treatment is required and then engaging in a quick conversation with the patient about the weather afterwards. It is equally important that the patient actually feel heard by the care worker. Patients are fully aware that a care worker's attention is a valuable commodity and not something that can be taken for granted, and that is why students in the health care sector are now being taught the importance of providing nonmedical attention, simply because of how scarce and valuable it is.

We are very good, it seems, at gauging whether or not we are getting the attention we crave, especially when we are feeling vulnerable. And we spend much of our time trying to figure out if our conversation partner is paying us enough attention, particularly because we know that we cannot take that attention for granted. A doctor who continues to stare at his or her computer screen during a consultation may come across as disinterested to the patient (and most patients are quick to spot this), even though the doctor may in fact need to focus all of his or her attention on arriving at the correct diagnosis, especially in complicated cases.

However, studies have shown that a correct diagnosis on its own is not enough to enable patients to feel satisfied with the care being provided; they also want to feel like they have been heard. Attention is, therefore, a crucial aspect of good health care.

Rapid Change

In the Netherlands, a clip from a documentary made by Frans Bromet in 1998 has become an internet hit twenty years after it was first filmed. The clip features the responses of a number of people on the street when asked if they have a cell phone or not. Most of them answer "no," and many cannot suppress a smile or a laugh when doing so. "Why would I want a cell phone? I have an answering machine at home!"; "I wouldn't like to have people ringing me all day"; and "Imagine people being able to ring you when you're on your bike!" A lot has changed since then. Wherever you look these days you see people staring at the screens of their smartphones. Whether we are waiting for the train or for the traffic lights to go green, we cannot tear ourselves away from our cell phones. It's actually a lot quieter on the trams in Amsterdam these days than it used to be. Whereas the youth of yesteryear would fill the carriages with their loud chatter, today's generation tends to sit or stand in silence staring at their smartphones. And to make matters worse, the last thing that many of us do at night before going to sleep (and the first thing we do in the morning) is check our social media.

Bromet's documentary is only twenty years old, but it is a good reminder that the mobile telephone and social media are relatively recent inventions and ones that we are still getting used to. The amount of new information that comes our way every day has increased dramatically over the past few decades. Scientists have estimated that people living in the United States took in five times more information each day in 2011 than they did in 1986. That is the equivalent of 175 newspapers of data per person, per day. The cell phones we carry around in our pockets have more processing power than the command center of the Apollo space missions had at its disposal. The increased availability of new information means that we now have to continually make more choices about what to select and what to ignore; choices that we must then stick to in order to be able to concentrate. For the purposes of this book, *concentration* is defined here as the action of maintaining one's attention on a particular task for a certain period of time without becoming distracted.

In this book, you will not only read useful tips on how to improve your concentration, but I will also explain how concentration works in the brain and discuss the latest scientific findings on this important topic. In the end, knowing how concentration works is probably the best way to improve your own concentration. The first step towards improved concentration is to appreciate the value of attention and realize that concentration should never be taken for granted, especially in today's busy world.

Concentration is very useful, but it can be easily disrupted by all the other information fighting for your attention (apps, external stimuli). So, at moments that require intense concentration, it is best to arrange your immediate environment in such a way that disturbance and distraction can be kept to a minimum. We might associate all the notifications we receive on our cell phones with potential rewards, but remember that conditioned relationships tend to become extinct over time. All you need to do is to change a few of your phone's settings and you will stop receiving notifications from social media. It may take a little getting used to at first, but the results can be very rewarding. You can also minimize the amount of distraction at your work not only by turning off the notifications on your computer but also by ensuring that there are fewer potential distractions around you. Even Pavlov's dogs stopped slobbering when the association had faded. If they can do it, so can you.

1
Why Is It Difficult to Concentrate?

I always leave my keys in a place where I will see them when I am walking out the door, mostly so that I don't end up locking myself out of the house. And when I have a letter that needs to be mailed, I usually put it somewhere obvious in the hope that I won't forget it when I'm packing my bag for work. I don't carry a handkerchief around with me, but, like many other people, I am in the habit of tying metaphorical knots in things to remind me of stuff I am otherwise likely to forget. We often use our external world to store information. A waiter in a restaurant can store a complicated drink order in an "external memory" by laying out the glasses behind the bar as soon as the order has been taken. He will then be able to recall which drinks he needs to pour without having to store the information in his own internal memory. It also allows him to take a new order without forgetting the previous one.

The benefit of using the external world as storage is that we don't have to burden our brain's "internal" memory. We use external memory banks to save much of the information we need to remember. We fill our diaries with appointments, write countless to-do lists, and leave reminders all around the house for things we are afraid we might forget. Most people are unable to recall more than two telephone numbers from memory because all of their important numbers are stored (relatively) safely in the memory of the smartphones. In fact, in many situations we prefer to rely on our external memory instead of our internal one. This is because our internal memory uses up a lot of energy, has a limited capacity, and is often unreliable.

The idea that our memory is not confined to our brain alone represents a relatively new approach to understanding cognitive functions like memory. Traditionally, experimental psychology used to focus exclusively on processes "inside" the brain (in the belief that we are our brains). But thanks to a new school of thought within cognitive science known as *embodied cognition*, we have since discovered that experimental psychology provides us with a limited view of our cognitive functions. The theory is that because our body is inextricably linked to our external environment, we cannot regard cognition as something separate from the way in which our body interacts with our surroundings.

For example, we are able to remember more of a story when we have physically acted it out. In an experiment, different groups of test subjects were instructed to read a story without knowing that they would be asked to recall the details later on. One of the groups was told to simply read the story, another group was asked to provide written answers to a number of questions about the story, and a third group discussed those questions with other test subjects. There was also a group of test subjects that was instructed to act out the story. All of the groups were subsequently asked to take a test related to the story. The results showed that the group that had acted out the story was able to recall more details, thereby demonstrating that the use of our physical body can lead to a more efficient memory. This also explains why taking notes helps you to remember more from the classes and lectures you attend, even if you throw away your notes afterwards. Writing is a physical act, after all. Furthermore, taking notes with pen and paper is also more beneficial than using a laptop, as the motor skills required for writing are far more complex than those used when typing on a keyboard.

How Does Information Access the Working Memory?

The short-term retention of information requires the use of our working memory, the part of the brain that is responsible for executing complicated tasks. You could compare the working memory to the tools and materials you lay out on a workbench. You might have a toolbox full of different tools and a whole shed full of materials, but you can only work with the tools and materials that you have before you on your workbench. You also need to keep the amount of stuff on your workbench to a minimum; otherwise, you will be unable to carry out your work in an efficient manner. This is why the choice as to which information you will place in your working memory is very important. Your working memory is crucial to your concentration because you use it to execute a specific task for a specific period of time. Therefore, the key to facilitating good concentration (or focus) is to know how your working memory works. I will first discuss the basic functioning of the working memory before going on to provide some concrete advice.

Roughly speaking, there are two ways in which information can access the working memory: from the outside world (through your senses) and from the internal world (when you think of something). However, not all of the information that your senses pick up from the external world ends up in your working memory. Which is a good thing, because otherwise everything you perceive would end up fighting for space on your workbench, and you would find yourself dealing with so much information that you wouldn't be able to think about anything other than what you pick up from the external world. To prevent this from happening, we select the most important information and focus our attention on it. Attention is the gate to our working memory: only the information upon which we focus our attention can access the working memory. Have a look at figure 1.1.

The first port of call for information picked up by your senses from the external world is the "sensory memory." Only a small portion of this information ever ends up in

Figure 1.1

the working memory. The sensory memory is known as the *iconic memory* when it relates to the visual system. The iconic memory is an extremely short-term memory that creates and retains detailed images of our world for the duration of only a few milliseconds. It acts as a kind of buffer for the images that are projected onto the retina. In the case of auditory information, we refer to what is known as the *echoic memory*. It remembers an echo of the sounds you have just heard. The sensory memory is a fascinating component of our memory system, primarily because we don't really have any direct access to it. It is possible, however, to catch a very fleeting glimpse of the iconic memory. If you light a sparkler in a dark room and wave it about, it will leave a kind of afterimage at the spot where it has just been, like a kind of tail. This is actually the content of the iconic memory that has not yet been overwritten by new information, given that in the dark there is nothing else for us to see. Of course, this doesn't happen very often because we usually find ourselves in well-lit environments, even at night, meaning that our iconic memory is constantly being overwritten.

In 1960 the cognitive psychologist George Sperling undertook a study of the capacity of this preliminary and very brief memory system to retain visual information. Test subjects were shown a number of letters (see below) on a screen for the extremely short duration of fifty milliseconds. He then asked them how many letters they had seen. As the letters had been displayed only for a very short time, the test subjects only managed to identify four or five letters at most. Sperling then came up with a trick to help the test subjects. He instructed them to look at the letters again and listen to a tone that would sound immediately after the letters had disappeared. When they heard a high tone they were to say which letters they had seen in the top row, in the case of a normal tone they were to pay attention to the middle row, and when they heard a low tone they were to focus their attention on the bottom row. This means that the test subjects were able to focus on a specific row in their iconic memory. Remember, the letters had already vanished by the time the tone sounded, meaning that they had to retrieve the information from their memory. And, amazingly, it worked! The test subjects were able to recall up to three letters per row when they focused their attention on one specific row in their iconic memory, indicating that they had somehow gained access to their iconic memory system. This means that the capacity of the iconic memory is far greater than that of the working memory, but also that we have very limited access to it. In the test, the tone had to be sounded immediately after the letters were shown. After only one second, the information had already disappeared from the subjects' memories and their performance did not improve, regardless of whether the tone was sounded or not.

Figure 1.2
An example from the Sperling iconic memory experiment

Most of us probably associate the word "memory" with things that happened a long time ago, but memory actually kicks in immediately after information has been picked up by one of our senses. No one knows the exact purpose of our sensory memory. Some scientists think it allows us to create a first impression of the world around us without having to establish all the details. The problem is that it can only be studied using the rather odd trick that Sperling came up with. What we do know is that attention does the job of ensuring that a portion of the information from our sensory memory ends up in our working memory. This is the moment when the information is laid out on the workbench so that our brain can think about it and perform computations on it. The information can be retained in the working memory for an extended period of time, as long as you keep your mind on it.

Figure 1.1 shows that information can also gain access to the working memory through the long-term memory, and thankfully so, because otherwise we would only ever be able to contemplate the information that we receive through our senses. The long-term memory is a collective term for all of the information that we can gather and store in our memory for a protracted period of time. Unlike the working memory, the capacity of the long-term memory is more or less infinite, although it is impossible to say exactly how much we are able to store in this memory system.

Our long-term memory contains all of our personal memories from the past, as well as the knowledge we have accrued over the course of our lifetime. Keeping information active in our working memory for a longer period of time allows us to store that information in the long-term memory. For example, retaining a telephone number for long enough in your working memory (by repeating it over and over or thinking about it constantly) will enable you to store it in your long-term memory. The curious thing about long-term memory, however, is that you can never be sure whether you have stored the information successfully or not. You may not even be able to access certain information in your long-term memory even though you know it is there. Information can only be retrieved by issuing the right instructions. We all know how frustrating it is not to be able to think of someone's name, all because we are not giving our long-term memory the correct instructions. We usually try to locate the information by thinking of things that we associate with that person. This can help to activate the memory. Conversely, the whiff of a particular fragrance can make you recall a memory you didn't think you had saved. Information that has been acquired within a specific context can be more easily retrieved within the same context. For example, if you study for an exam in the same room as where the exam is to be held, you have a better chance of achieving a higher score than if you study for the exam in a different place. Recalling the location where you learned something can help to retrieve specific information from your long-term memory.

When you are concentrating on a certain task, for example reading an email, all of the information related to that task is stored temporarily in your working memory. At the same time, new information related to the email continues to access your working memory. However, irrelevant information can also access your working memory: if a pop-up window appears on the screen while you are reading the email, your attention will automatically be drawn to this new information. This poses a threat to your concentration because the irrelevant pop-up information can cause other information to be pushed out of your working memory. In fact, you may even find that you no longer know exactly what you were doing and that you have lost your concentration entirely.

Retaining Information in the Working Memory

How often have you walked around for days with a letter in your bag that you keep meaning to mail? You keep forgetting to do so despite passing a mailbox several times a day on your way to and from home. This happens because if you do not repeat the task to yourself over and over it simply disappears from your working memory. The only way of making sure you will mail the letter is to retain the task in your working memory by reminding yourself repeatedly. This is doable, of course, but it is extremely tedious. To ensure that the task will not vanish from your working memory you must not think of anything else while reminding yourself not to forget to mail the letter. A different thought or something in the external world that catches your attention will encroach upon your working memory and may even push the task of posting the letter out of your memory altogether, with no guarantee of it ever returning. Worse still, you cannot prevent this from happening because you never know in advance exactly what will happen during your day or whom you might bump into.

Imagine you are asked to do a quick calculation. For most of us, it's not too difficult to find the answer to something like 15×13. We simply divide it up into a series of steps: first multiply 10 by 13 and 5 by 13 before adding the two together. However, more complicated calculations do present a problem. For example, most people would find it very difficult, if not impossible, to work out the solution to 323×144 using only their working memory. This is because the working memory can only hold a certain amount of information at any one time. We refer to each piece of information currently occupying the working memory as an "item." On average, the working memory can hold around six of these items, although it must be pointed out that remembering two items is a lot more difficult than remembering just one. And that is why solving this problem is very difficult for most of us: if you use the same method as you did for the less complicated problem above, you have to be able to remember 323×100 as well as the number 44. A complicated math problem involves a lot more steps and the numbers you need to remember are made up of a larger number of items.

There are major differences in the amount of information that different people are able to retain in their working memory (i.e., the "capacity" of the working memory). In many of the studies on the functioning of the working memory, a differentiation is made between test subjects with a small capacity and those with a large capacity. This capacity provides an accurate indication of how a person's other important cognitive functions are performing: it can predict the extent to which a person may benefit from a memory-training program after they have suffered brain damage, the linguistic abilities of patients with schizophrenia, or how well a person can retain information over a long period of time, among other things. The relationship is always one-way: the greater the working memory capacity, the better a person will be able to perform complicated tasks, such as solving math problems and taking an IQ test. Even at a very young age, the working memory can be an accurate gauge of how well people will be able to solve problems in a logical manner later on in life.

Fortunately, it is possible to train your working memory so as to increase its capacity. For instance, people who are known to have a smaller capacity, such as children with ADHD or a developmental delay, can enlarge the capacity of their working memory by playing memory games. This can also lead to improved performance on more complicated tasks that we know are related to the working memory. However, these improvements are only visible for the group as a whole, so it does not necessarily mean that training works for everyone with a low working memory capacity.

In addition to training, we can also increase the capacity of our working memory by using the external world as a memory system. When we are trying to solve a difficult mathematical equation, we can write down the results of the different steps so that we don't have to store that information internally but are still able to access it immediately when we need it. This allows us to use our working memory to solve subsequent steps without having to remember the results of the previous steps as well. Of course, you could just grab a calculator and punch in the sum, which is also a way of using an external working memory.

To retain information in the working memory, it has to be repeated over and over. In a classic experiment in 1959, the psychologists Lloyd Peterson and Margaret Peterson asked test subjects to remember a number of letters, such as CSP, WKL, and SRP. After the letters had been displayed very briefly and had vanished from the screen, a tone was sounded, signaling that the test subjects had to name the letters they had just seen. Under these straightforward conditions, it did not make any difference how much time elapsed between the letters vanishing and the tone being sounded—all of the test subjects were able to recall the letters. The results were different, however, under more difficult conditions in which, after the letters had vanished, the test subjects were shown a number and asked to sequentially subtract the number 3 from that number while simultaneously trying to remember the letters. For example, they were shown the number 456 and had to perform the subtractions in steps, giving the numbers 453, 450, 447, etc. You have probably already guessed that in this case it made a major difference how much time elapsed before the tone was sounded when the letters had vanished. Performing the subtractions encroached upon the capacity of the test subjects' working memory. After ten seconds of sequential subtractions they were able to recall very few of the letters, and after eighteen seconds they were only able to identify 5 percent of the letters correctly. This is because you need your working memory to be able to perform subtractions. If you don't believe me, try telling a story while solving a mathematical equation at the same time. It simply can't be done, because both tasks require the use of our working memory and that memory can only perform one task at a time.

The repetition of information in the working memory helps to ensure that the information will not be forgotten and improves the chances of it becoming stored in the long-term memory. Having to perform even a relatively simple calculation can prevent other information from being retained in the working memory, in which case the information may be lost forever from the brain. That is why it is crucial to avoid all distraction during periods of high concentration. The information that is essential to the task you are performing can disappear in the blink of an eye if you stop thinking about it.

Which Information Can You Store in Your Working Memory?

Time for a quick experiment to test your working memory. Look quickly at the letters below, close your eyes, and then try to recall the letters:

USANASANATO

That wasn't too difficult, was it? And you even managed to fit eleven letters into your working memory! Significantly more than the six items I mentioned above. But let's not get carried away. It is highly likely that you have just used a well-known memory aid. You probably saw that the eleven letters are made up of three acronyms—USA, NASA and NATO—and grouped the individual letters together so that instead of having to remember eleven letters you only had to recall three sets of letters. This is known as *chunking*: grouping individual letters, numbers or images together so that you can retain more information in your working memory than when you try to remember all of the separate items.

This shows that even a complex piece of information can be retained in our working memory as a single item. You are probably familiar with this phenomenon from trying to remember your telephone number or the number of your bank account. The IBAN bank numbering system that was recently introduced in Europe uses numbers that at first glance are not easy to learn off by heart. But if you know how the IBAN number is put together, it becomes an awful lot easier. For instance, the letters constitute a bank code, which is directly related to the name of the bank. Of course, you can try to commit the eighteen individual numbers and letters that make up your IBAN to your working memory, but you will save yourself a lot of heartache by chunking the bank code (i.e., the name of the bank) and storing it as one single item, thereby saving space in your working memory.

Many of the techniques used to increase the capacity of the working memory are based on the effective chunking of information and its effectiveness depends on the familiarity of the information in question, which can vary from person to person. A good example is that of the long-distance runner who taught himself to remember numbers in the form of race finishing times. This enabled him to store up to eighty digits in his working memory! Unfortunately, this technique is of little use to someone who is not well acquainted with record race times.

Up to now, the examples we have used have focused primarily on how to store digits and characters in the working memory. But our minds are often preoccupied with visual imagery or songs instead. In order to establish what we can fit into our working memory, and subsequently focus our attention on, it is important to first identify the different components of the working memory. Possessing that knowledge makes it easier to see which tasks we can and cannot perform simultaneously, given that each individual task demands the attention of a different part of our working memory.

A leading pioneer in the study of working memory is Alan Baddeley, a lecturer at the University of York whose career was launched back in the 1950s with a study of how to memorize postal codes. At the time, the British postal service was attempting to modernize the postal system by introducing postal codes across the nation, and we have Baddeley to thank for the postal codes still in use today in Great Britain. Each code consists of six letters and digits, a number that is based on the capacity of the working memory. To make it easier for people to remember their code, Baddeley suggested that each code should begin with the first letters of the name of the town or city in question (such as BA for the city of Bath). This part of the code was easy to remember (thanks to chunking) and had the added bonus of helping to ensure that a person's letter still ended up at the correct destination if they made a mistake with the second, more difficult part of the postal code. This second part consists of letters and digits, with the digits being written first so as to avoid possible confusion with the first part of the code (BA 27AY, for example). Baddeley's method proved very successful, and a survey recently carried out by the postal service revealed that 92 percent of the population still find it easier to remember their postal code than their PIN or the date of their wedding anniversary. Even though a PIN or wedding anniversary contains fewer characters than a postal code, it is still easier to remember the postal code because of the chunking of information and the efficient sequence of the letters and digits. The American postal code is less forgiving because the only clue it provides with regard to location is in the abbreviation of the state name.

By the time he completed his work on postal codes, Baddeley was well and truly hooked on the subject of memory and was eventually responsible for devising what still is the most influential model of the working memory. In his model, the working memory is made up of three components. The main component is the *central executive*, which is responsible for initiating and checking cognitive processes, and which issues commands like the captain of a ship. It can steer the brain's linguistic functions in order to decipher a piece of text, or it can tell the eyes what to look at. The central executive makes use of the working memory's two storage systems: the *visuospatial sketchpad* and the *phonological loop.*

The phonological loop's job is to retain auditory and linguistic information, such as sounds and spoken text. If someone has just told you their telephone number, you can

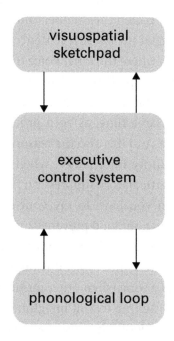

Figure 1.3
The working memory, a central executive with two storage systems: one for spatial information (the visuospatial sketchpad) and one for auditory and linguistic information (the phonological loop)

retain it in the loop by repeating the number to yourself and hearing the information over and over again. (When reading the next few sentences try to remember the numbers 1, 5, and 8). This system is also active when you are reading. We transform the information we read into silent speech that is then loaded into the phonological loop. Processing the information in the brain's linguistic regions enables us to understand the text. Each word that you read is held in the working memory until you have processed its meaning. (Wasn't easy, was it, to remember the numbers? You have probably either forgotten the numbers altogether or failed to grasp the meaning of the last few sentences.)

The loop is also where the inner voice that we use to think and to reason ends up. This explains why it is impossible to read and understand a piece of text and think of something else at the same time, as both processes make use of the same capacity. And it is also the reason why it is difficult to remember auditory information when you are speaking out loud (a.k.a. "articulatory suppression") or to recall someone's name when you have to say your own name at the same time. Speaking out loud prevents auditory information from being repeated in the loop, resulting in the loss of that information. For the same reason it is impossible to remember the melody of a song when other music is playing simultaneously (both listening to and imagining music make use of the phonological loop). It is also a good idea to repeat the name of someone you are meeting for the first time before saying your own name.

The Capacity of the Phonological Loop

Given that information in the working memory has to be actively repeated, it is a lot more difficult to recall a long list of words in the phonological loop than a short one.

Try it for yourself:

List 1 = party, joke, dog, star, walk, yellow

List 2 = coincidence, cheeseburger, telephone, instructor, mouthwatering, congestion

You will have noticed that the first list is a lot easier to remember than the second one. When repeating the longer words, there is a greater chance of one of the other words being erased from the working memory. After all, the more time that elapses since you last repeated a word, the more likely you are to forget it. The phonological loop does not have a fixed capacity for the number of words it can hold. Instead, that capacity depends on how quickly a word can be repeated.

The phonological loop is very important in our daily lives because thinking and communicating are linguistic processes. We use the loop continually: when speaking, while reading, and even when you are talking to yourself. We often use internal speech to calm ourselves down, like when we become angry about something, for example. It allows us to control our impulses and our heart rate. Where your first reaction to an antisocial driver may be to call him every name under the sun, your internal voice can prevent you from actually doing so. The loop does have a limited capacity, however, meaning that you cannot think of several things at the same time and that some of its content may be lost when someone suddenly starts talking to you.

In addition to the phonological loop, the working memory also has another storage system: the visuospatial sketchpad. This is where visual information and information about locations in the world are stored. You also use it when you are imagining something. For example, when you think of a person's face or the layout of the house where you were born, that information is retrieved from the long-term memory and retained in the visuospatial sketchpad. If you go looking for a red book on a bookshelf, you activate the color red in the visuospatial sketchpad so that all of the red objects around you immediately attract your attention. This means that you can use the contents of the visuospatial sketchpad to search your environment. If you can't find your keys, it helps to visualize those keys so that you can search for them more efficiently. And when you want to be able to remember where you left your keys, you save the location in the visuospatial sketchpad. It's a bit like sticking a drawing pin into a spot on a map of the world. To remember the location of your keys, you stick a pin in that location and store the memory in the visuospatial sketchpad.

In addition to storing visual information, you can also "edit" that information in the visuospatial sketchpad. For example, you can imagine an object you have never seen before in real life and give it another color or spin it around. Take a look at the illustration below. Which of the four numbered figures is a rotated version of the first figure on the left? To be able to figure this out you have to execute a mental rotation of the other figures in your visuospatial sketchpad. Studies have shown that the amount of time this takes

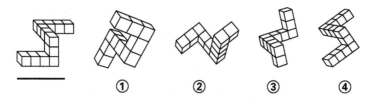

Figure 1.4

is influenced by the number of steps required to arrive at the answer. This means that you have to rotate the object in your head the same way as you would if you were holding it in your hands. The correct answer, by the way, is 4.

Research has shown that athletes and musicians are a lot better than most of us at this kind of mental rotation. In tests, they react much faster and make fewer mistakes than other test subjects. You can train yourself to become better at a task like mentally rotating an object. The effect of physical exercise was demonstrated by a study that compared the results of two mental rotation tests: one carried out before test subjects underwent a physical training session, and another conducted afterwards. Competing in sports or playing music at the highest level requires you to be able to imagine not only your own actions but also those of your opponent or teacher. This kind of mental training allows you to make your visuospatial sketchpad even more efficient and improves your ability to perform tasks like mental rotation.

And, sorry to have to say it, but the often-cited claim that men are better at solving spatial problems than women is, in fact, true. Striking differences have been measured even in three-month-old babies. When researching this phenomenon, scientists studied how long babies looked at a certain object on a screen and then concluded that when a baby looks at an object for a longer period of time it means that they find that object more interesting than the other objects on the screen. Baby boys tend to look longer at the rotated variant of an object and spot it more quickly than baby girls do. This does not necessarily mean that boys are better at this at a young age, but it does demonstrate that there are differences in the way babies perceive rotated objects. It must also be said that this does not mean that these differences are congenital. They may also originate from boys being given different toys than girls from a very young age, toys that are more likely to allow them to develop their skills at mental rotation. Think, for example, of Lego, a toy that typically requires mental rotation skills. So, the next time you're buying a gift for a young girl, you might want to explore the Lego section—although there are plenty of other toys that require mental rotation, such as the humble building block.

We can chunk information in the visuospatial sketchpad the same way we chunk letters into a word. After all, a visual object can have a variety of colors and shapes. In this case, you remember not only the individual colors and shapes as separate items, but the complex figure as a whole counts as a single item as well. It appears that the visuospatial sketchpad is able to remember four of these complex figures without too much trouble. Using the same capacity, you can remember four separate colors or shapes or four complex objects of different shapes and colors. The visuospatial sketchpad is therefore able to remember whole objects, regardless of how complex they are. And this is a good thing, too, because in our daily lives, our visual world is mainly made up of complex figures and not of simple, colorless objects, as is often the case in many of our experiments.

The Role of Attention in the Working Memory

The two storage systems described above work more or less independently of each other. We have already seen that spoken language can prevent linguistic information from being retained in the phonological loop, but the loop cannot be interrupted by mentally rotating an object. Mental rotation is not a linguistic task and therefore has almost no influence on the processes in the phonological loop. The central executive ensures that attention is evenly split between the two storage systems.

Imagine you are driving a car and your passenger is giving you instructions about the route ahead. These instructions come in the form of verbal information, which you must then convert into a visual image of the route. At that moment, you are using both storage systems, and it is the working memory's job to ensure that enough attention is paid to both processes so that all of the information can be retained. If you pay too much attention to the visual aspect, you will probably end up saying, "Hang on, what did you say? I wasn't paying attention." At that moment you were probably paying too little attention to the information in the phonological loop and consequently lost track of it. At the same time, the central executive has to make sure that you ignore all other incoming information. For example, the radio might be switched on and relaying other verbal information, like traffic reports. The central executive acts as a kind of umpire who makes sure that you ignore this information and focus your attention on the information you need instead.

Being able to focus your attention on a specific task in the working memory is crucial to good concentration. In theoretical attention models, this kind of attention is referred to as *executive attention*, meaning that it is required to be able to execute a certain activity. Other terms used to describe the functions of the central executive include "cognitive control" and "executive function." They all mean more or less the same thing: concentrating on a task and executing that task successfully by focusing on the relevant instructions for a certain period of time and monitoring and adjusting your own behavior where necessary.

Imagine you are playing a game of tennis and you have a secret weapon up your sleeve. Each time you are faced with an important point in a game, you play a drop shot—a short shot just over the net. In the first set, it works every single time. Your opponent fails to anticipate the drop shots and is caught off guard time and time again. In the second set, however, your opponent has cottoned on to your trick and is ready to pounce each time you attempt a drop shot. You lose the second set and realize that you will need to adjust your strategy. This is the moment when you need to change the rules governing your working memory and make other choices—for example, to aim more of your shots at the opponent's baseline. This is a job for your executive attention: adjusting your strategy. If you fail to focus on this task, you will simply carry on playing on autopilot without realizing that you are making the same mistake over and over again. You need concentration to be able to evaluate your actions and adjust them where required.

People who suffer from problems with their working memory (for example, as a result of brain damage) are less able to adjust a successful strategy when that strategy has long since stopped working effectively. This is referred to as *perseveration* and it reveals a very important function of our working memory: it enables us to interact with the world around us in a flexible manner. This flexibility is often measured using the *Wisconsin card sorting task*, in which subjects are asked to sort a pile of cards following a specific rule. As you can see below, the cards have a number of characteristics that enable them to be sorted: color (in this case, shades of gray), shape, and quantity. There are four separate piles of cards, and the loose cards can be sorted using one of the three characteristics. If color is the determining characteristic, the card must be placed on the first pile. If it's shape, the card goes on the fourth pile. And if the determining characteristic is quantity, the card is added to the second pile.

The rule for sorting changes after a certain number of cards. If you have been sorting on the basis of color for a long period of time, you are told to switch to sorting on the basis of quantity—just like when you stopped playing drop

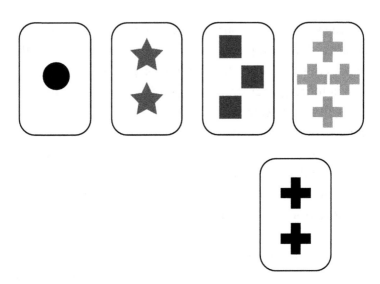

Figure 1.5
An example of the Wisconsin card sorting task

shots at crucial moments in the third set of your game of tennis. This proves very problematic for patients who suffer from perseveration. They continue to sort the cards according to the old rule despite being told to follow the new one, even after they have received feedback on their performance and know exactly what they are doing wrong.

Children and the elderly have more difficulty performing this task than young adults. This is because the prefrontal cortex is largely responsible for our working memory and that part of the brain is not only the last to reach maturity but also the first to start deteriorating with age. The prefrontal cortex dictates our ability to control whatever task it is we are performing. You could regard it as the control center of our brain. There are strong neural links between the prefrontal cortex and the other parts of the brain and that is why it contains the unique kind of infrastructure required to control and steer the activities in the rest of the brain. These links are a lot stronger in the human brain than they are in the brains of other animals, which explains why humans are better at planning, organizing, analyzing, and reasoning than any other animal.

Good concentration allows us to interact with our surroundings in a flexible manner. Each situation requires a different set of rules. For example, you speak to your boss in a different way than you do to your neighbor (provided they are two different people, of course). And you need to use a lot of strength to open a heavy door, but using the same amount of strength to open a much lighter door can have painful consequences. It is possible to arrive safely at work even if your mind is not entirely on the road while driving because you travel the same route every day. However, it can be a problem when your intention is to drive to the gym on a Saturday morning only to suddenly find yourself parked outside your office. When my department moved to new premises, I found myself walking into our old building on a regular basis just because I had not been concentrating on the route and walking to work on autopilot.

Concentration ensures that you don't become distracted when performing a certain activity. If you can make sure you pay no attention to irrelevant matters, you will be able to carry out your task without interruption. You might think that the ideal situation is never allowing yourself to become distracted from whatever you are doing, but that is not the case. From an evolutionary perspective, it is, in fact, extremely dangerous not to allow yourself to become distracted during moments of concentration. Long ago, humans needed to be alert at all times in order to be able to spot potential predators. And this is why we still get a fright when someone taps us unexpectedly on the shoulder while we are concentrating hard on something. This reflex sparks a flight response and enables us to run from danger if we need to.

Our brain is constantly monitoring our surroundings and, in the case of danger, will provide us with a signal that we should immediately stop what we are doing. To do so, it interrupts the working memory by using what is referred to in cognitive neuroscience as a "circuit breaker," which allows the brain to initiate an automatic shutdown. While you are sitting back reading this book, this system is permanently monitoring your safety. The system is located primarily in the right-hand side of the brain in the ventral frontoparietal network, a cluster of brain parts whose job it is to react to the presence of unusual objects (even when they are situated beyond the focus of our attention). We have little control over this system, and that is exactly the way it should be. After all, in dangerous situations we do not have the time to decide whether or not something deserves our attention; the important thing is that we immediately stop whatever we are doing. And run!

There is a classic distinction between two different types of attention: voluntary attention, which is based on the tasks that are currently being performed in the working memory (the book you are reading, the radio show you are listening to) and automatic, reflexive attention over which you have no control and that can cause information to be pushed into the working memory. Reflexive attention is important because it enables you to keep an eye on your surroundings during periods of high concentration, but it can also interrupt your concentration. However, not all stimuli that break your concentration are necessarily warning signs. It is important, therefore, that you do all you can to avoid irrelevant stimuli when you want to concentrate on something and not to surround yourself with devices that may cause unwelcome distraction. You probably understand by now why it is extremely beneficial to your concentration to turn off all the notifications on your cell phone and computer.

When Your Working Memory Is Too Full to Allow Good Concentration

In December 2016, a video was posted online that achieved millions of views in a very short space of time. At first glance, it appeared to be nothing more than an innocent love story shot at a typical American high school; a story in which two students leave a series of messages for each other on a school desk. They eventually meet for real near the end of the video. In the last shot, however, we see another student walk into the room brandishing a gun. It is an obvious reference to the mass shootings that have plagued schools in the United States. The video was made by Sandy Hook Promise, a non-profit organization whose goal is to prevent shootings like the one that took place at Sandy Hook Elementary School in 2012 from happening again.

One of the reasons the video is so gripping is that at the end it reveals a number of signals you might have noticed if you hadn't devoted all of your attention to following the love story. We are shown examples of the deviant behavior of the student who appears at the end with a gun. He dismisses all contact with other students, makes gun-toting gestures, and searches the internet for information about guns. All of this happens in full view of everyone, but no one notices. And therein lies the message the video wishes to get across: pay attention to what goes on around you; someone might just be preparing a mass shooting right under your nose. As the Sandy Hook Promise slogan says, "Gun violence is pre-ventable when you know the signs."

Despite the video's honorable intentions, it actually conveys the wrong message. It asks the impossible of the brain, because we are only able to focus our attention on one thing at a time. The danger is that, after watching the video, people will feel guilty about not having spotted things that were being played out in the background, however subtle they may have been. Imagine if we had to continually focus our attention on the things going on in the background of our lives. If that were the case, we would never be able to concentrate on the stuff that really matters. This is exactly what people who suffer from post-traumatic stress disorder (PTSD) have to deal with. People with PTSD have difficulty concentrating because they are constantly and involuntarily scouring their surroundings for signs of danger. This means that part of their working memory is always active, which results in major concentration problems.

PTSD is the result of severely stressful situations involving a traumatic and often life-threatening event, sometimes paired with serious physical injury. Research into PTSD was initially driven primarily by the war in Vietnam, although stories of soldiers suffering from PTSD can be found as far back as ancient times. PTSD is usually associated with the military, but it can be the result of abuse or a serious accident. Patients tend to suffer badly from nightmares and exhibit extremely high levels of nervous tension at the slightest provocation. A typical example is the nervous reaction of a patient to the sound of a door slamming. The brain has to operate under permanent stress and this has serious consequences for the working memory. Brain scans reveal anomalies in the brain activity of PTSD patients in the regions that are responsible for the working memory. This affects their level of control over everyday activities, and sometimes these patients appear to be in a constant state of multitasking.

Even those who don't suffer from PTSD acknowledge the effect anxiety can have on concentration. Have you ever tried reading a book while you were feeling scared? Impossible. Anxiety forces your working memory to give priority to the constant monitoring of your surroundings. Every single detail is subjected to inspection, the kind of behavior that is recommended in the Sandy Hook Promise video.

We all worry about things from time to time. But if you worry too much about something, like taking an exam, you can freeze up completely. In that case, the worry takes up so much of the capacity of your working memory that you are unable to perform to the best of your ability (known as *choking*). Given the limited capacity of the working memory, it is not surprising that when you overburden your working memory it affects your performance, especially when you are carrying out a complicated activity. It also means that if you can alleviate the anxiety, your performance will improve. An excellent experiment carried out to demonstrate this made use of two groups of test subjects. One of the groups had to solve a difficult mathematical equation immediately after spending ten minutes sitting still and doing nothing. This group had all the time in the world to worry about the impending task, even more so because they knew that the potential reward depended on their performance in the test. The other group was not required to sit still for ten minutes but was instead given the opportunity to write down their thoughts and feelings about the upcoming test. Even though both groups had performed equally well at an earlier test with no conditions attached, the group that had to sit still for ten minutes made significantly more mistakes in the second test than the group that was allowed to put their thoughts down on paper.

Even more interesting perhaps is the fact that the second group performed better at the second test than they did at the first one. It appears that being able to write down your thoughts and concerns before a test or performance can help you to declutter and even sharpen your working memory. This may be related to the fact that the physical world can serve as an external memory. In this case, the worries are removed from your head and stored in the external memory represented by the sheet of paper, thereby freeing up the required capacity in the working memory.

There are limits to how much you can demand of your working memory; overburdening it tends to lead to mistakes. When the working memory is filled with information, there is little capacity left over for other activities. If you tell test subjects that they will be filmed while they are performing a difficult task and evaluated afterwards, they will perform worse. The thoughts associated with this information have to be processed in the working memory, which makes it even more difficult to carry out the already difficult task. This explains why coaches often tell their athletes to empty their minds before an important event. The limited capacity of the working memory and the vulnerability of information contained there means that it is never easy to maintain your concentration. So, if you want to be able to concentrate properly, you have to make optimal use of your working memory and avoid all unnecessary distractions.

2
When and When Not to Multitask

Brian Cullinan, an accountant at the prestigious PricewaterhouseCoopers firm, considered himself luckier than most in his job. After all, he was responsible for handing out the prizes at the annual Oscars ceremony. He and his colleague, Martha Ruiz, were tasked with ensuring that the votes were counted correctly, that the envelopes containing the winners' names were in order, and that the right envelopes were handed to the right people on the night. In an interview with the BBC one week before the awards ceremony, they had proudly boasted how the entire voting procedure was completely watertight. The seven thousand votes of the members of the Academy of Motion Picture Arts and Sciences were counted multiple times and the results sealed in envelopes before being locked in a safe. They always made two sets of envelopes, one for Brian and one for Martha, who then traveled separately and under heavy security to the awards ceremony. Safe as safe can be. At the awards ceremony they stood on opposite sides of the stage so that they would be able to hand the right envelope to the presenter just before he or she came on stage, regardless of which side they chose to make their entrance. All that either Brian or Martha had to do was to make sure they disposed of their copy of the envelope containing the names for each award when the other copy had been handed over on the opposite side of the stage. What could possibly go wrong?

The 2017 Oscars provided one of the most embarrassing moments in the ceremony's long history. Martha had just handed over the envelope containing the name of the winner of the Best Actress award, Emma Stone for the movie *La La Land*. The next prize to be announced was the most important of the night: Best Picture. However, the presenters, Warren Beatty and Faye Dunaway, were handed the wrong envelope by Brian. He had forgotten to dispose of the envelope for the previous award and instead handed it to Beatty. He then opened the envelope on stage, hesitated for a moment when he saw what was written inside—"Emma Stone for *La La Land*"—before handing the envelope to Dunaway, who was eager to get on with it. It was her job, after all, to announce the name of the winner of Best Picture. She read out the name of the movie printed on the card, *La La Land*, and didn't notice the name Emma Stone written above it.

By the time the mistake was discovered, the makers of *La La Land* were already busy giving their acceptance speech. The long delay may have been because there was no contingency plan in place for dealing with a situation in which the wrong envelope was handed over. Brian Cullinan even confirmed this in an interview with the *Huffington Post* before the ceremony—the chances of something going wrong are so small that it wasn't considered necessary to make provisions for such a scenario. In the end, Martha and Brian had to walk onto the stage to sort out the mess, after which the audience were informed of the error and the Oscar was presented to the winning movie, *Moonlight*.

How could something like this happen? There is no shortage of theories, of course, but one fact in particular stands out. Brian is a fervent user of Twitter. In the months leading up to the ceremony, he posted regular tweets about how much he was looking forward to the night and also about the ins and outs of the awards procedure. And on the evening in question he posted a lot of tweets too, including a photo of Emma Stone backstage with her award. He deleted that tweet afterwards, but not before everyone realized that he had posted it just after Emma Stone had walked off stage with her award, the exact moment when Brian should have been disposing of the envelope for best actress and getting the next one ready in case the presenters planned to walk on to the stage from his side. Unfortunately for Brian, and everyone else for that matter, the presenters did choose his side of the stage. The whole episode was not just a major blunder on the part of the Academy, as Pricewaterhouse-Coopers was left with a lot of egg on its face too. A few days later, the firm issued a press release in which it accepted full responsibility for the incident.

The job Brian was tasked with may not seem very difficult, but neither is it one that can be carried out on autopilot. Taking care of the envelopes requires the use of the working memory, and when this task is combined with another task (such as typing a tweet), it means that the working memory has to carry out two tasks simultaneously. And therein lies the problem. Our brain is not capable of taking on two tasks at the same time when both require the use of the working memory. So if you believe you are able to do several things at once without any difficulty, you are more than likely the victim of the illusion of multitasking. What happens is that you switch so quickly between two tasks that it *appears* as if you are doing them simultaneously. This is not the case, however, and you can actually observe this in the brain. The two halves of the brain charged with executing the tasks become alternatively and not simultaneously active, meaning that the brain has to switch continuously between the two activities. Multitasking can therefore be more accurately described as switching between tasks as opposed to combining them.

But is switching between tasks actually all that easy? And is it possible to do so without any negative effects? Would Brian have made the same mistake if he hadn't grabbed his phone to take a photo of Emma Stone and share it with the world? To answer these questions, we need to look at the experiments that have been carried out on *task switching*. Experimental psychology is a wide-ranging field of study made up of lots of different groups of researchers, each one interested in a specific area of research. They all organize their own conferences and publish their own articles in their own journals. One of these groups spends all of its time studying task switching. And not without good reason, because task switching tells us an awful lot about how flexibly our brain is able interact with the world around us. We are constantly being asked to do something different, depending on the situation, like search for our wallet so that we can then find our commuter pass before looking for the right platform and making sure we don't bump into other people while running to catch our train.

In the lab, this switching is studied using more abstract kinds of tasks, but the principle remains the same. One example is the *number-letter task*, in which test subjects are asked to switch between categorizing numbers and letters. Before each task they are shown the word "number" or "letter," after which a number-letter combination appears on the screen (e.g., 2B or N3). When the task is related to numbers, the test subjects have to say as quickly as possible whether the number they see is odd or even. For the letter task they must say whether the letter is a vowel or a consonant. Sometimes they have to switch between tasks (e.g., first the letter task, then the number task), and sometimes the task is repeated (e.g., the number task twice in succession).

Scientists all over the world consistently make the same observations in their experiments. On average, test subjects are slower to react to a number-letter combination when they have to switch between two tasks than when the task is repeated, a fact that can have far-reaching consequences. There are always costs attached (in terms of reaction time and the number of mistakes) to switching between tasks that require the use of the working memory. It will take you longer to complete each task (or "switch costs"), and you will make more mistakes than when your attention is focused on a single task. From the previous chapter we know that all tasks that are not carried out automatically but that require your attention will draw on your working memory. The extent to which they do so depends on the task's level of difficulty: the costs are higher when you have just carried out a complicated task than when it is less difficult but still requires the use of the working memory. After all, it takes more time to tidy a very messy room than one that is relatively clean.

This explains why the switch costs are lower when test subjects know in advance which task they will have to carry out after the switch. It allows them to clear their working memory and prepare for the next task before it comes along. In one experiment, test subjects were asked to make a difficult jigsaw and were then interrupted during that task and told to switch to another, less complicated task. The test subjects' performance on this second task was considerably poorer when they had to switch to it straight away, compared to when they were first allowed to finish the jigsaw and "tidy it up" in their working memory. This explains why interruptions from outside are often more of a hindrance than when you make the switch yourself. When your work is interrupted, a remnant of it remains behind in the working memory because you have not been able to empty your working memory completely. And the larger the remnant that you carry over into your next task, the higher the switch costs will be.

Interestingly, these scientific studies never show up any differences between men and women in terms of their ability to switch between tasks. And yet, despite all the careful scientific analysis, the myth that women are better at multitasking than men still persists. No evidence to back up this claim has ever been found. However, it has been established that some people are better than others at switching between tasks. It simply costs them less time and effort, regardless of their gender.

Media Usage and Multitasking

We all know them (indeed you might be one of them yourself), those people who are always checking their email and surfing the net when they should be studying or concentrating on their work instead. Ninety-five percent of us spend an average of one-third of the day simultaneously keeping tabs on various (social) media, for example, checking Facebook on your smartphone while watching TV and keeping an eye on Twitter on your tablet at the same time. Researchers at Stanford University have discovered that people who spend a lot of their time on various media (e.g., on their smartphone or laptop) have higher switch costs than people who have a lower level of media usage. The level of multimedia usage is one of the factors that predict how efficiently the brain is able to switch between tasks. Furthermore, switching tasks was not the only area where heavy media users registered a poorer score. They were also more easily distracted by incoming information and performed worse at memory tasks in which they were asked to remember certain letters.

At first glance, these results may appear to be very surprising. After all, given that we spend so much of our time multitasking, you would be forgiven for thinking that our brains should be getting better and better at switching constantly between tasks. However, that is not the case. In fact, we tend to overestimate our ability to multitask. In experiments, test subjects have consistently proven to be very poor at gauging their capacity for multitasking and their subsequent performance. Young people generally believe that they are quite capable of using six or seven different media simultaneously, but that is simply impossible, regardless of your age.

Although it is clear that multitasking is not good for your concentration, there is no reason to believe that it causes any permanent damage. Based on the research into the media usage of heavy multitaskers, you could possibly conclude that switching too often between tasks is bad for your brain and makes it less efficient, but there is also another possibility: it may be that heavy multimedia usage has no permanent effect on the brain at all. Instead, it may be the other way around—that people with less efficient brains are more likely to use different media all at the same time. In other words, people who are more easily distracted may be more inclined to use multiple media simultaneously.

The more you think about this, the less unlikely it seems. If you find it difficult to ignore distractions, you will automatically be more inclined to multitask. The stimuli with which you are constantly bombarded, like the notifications you receive on your smartphone or the mere presence of your tablet next to you, will command your attention and distract you from whatever it is you are doing. And this happens more easily to people with a less efficient brain. There is no evidence that increasing multimedia usage has a permanently negative effect on people, but we do know that the increased opportunity for multitasking can form a significant challenge for people with a less efficient brain.

Multitasking on the Work Floor

How often do you find yourself doing several things at the same time at work? According to research, most of us are guilty of this. It depends, of course, on what kind of work you do, but very precise results have been produced for certain professions, particularly those that involve working in shared office spaces. In such a situation, the chance of you becoming distracted from your work is far greater than when you have an office to yourself. It turns out that task switching is often the result of employees being interrupted by others.

In 2005, researchers carried out an observational study of typical office workers (such as financial analysts and software developers). They used a stopwatch and a notepad to track the behavior and activities of employees over the course of their working day. The employees in question were all working on different projects simultaneously, meaning that they often had to do several things at once. The study revealed (after an observation period of 700 working hours) that, on average, the workers were interrupted every eleven minutes and that the interruption usually caused them to shift their attention to a different task. The interruptions included answering the telephone, questions from colleagues, and incoming email. When the interruption was not directly related to whatever they were originally doing, it took another twenty-five minutes before they returned to that task.

Not all of the switches were the result of outside interference. Employees were also inclined to switch tasks of their own accord. Later on in interviews, they indicated that their preference would be not to do this, but that they felt they were forced to because of the different priorities attached to the projects they were working on. This meant that instead of working for longer periods of time on a single project, they ended up doing the bare minimum across a wide range of projects. The workers had to constantly monitor what the most important tasks were at any given moment among all of the projects in which they were involved.

Another reason for switching between tasks is a lack of concentration. Many workers find it difficult to work on one task for a significant period of time and often end up making a phone call or checking their email after only ten minutes have elapsed. A study conducted by an Australian telecommunications firm revealed that most of its staff spent less than ten minutes working without interruption on a single task and that the average was as low as three minutes. Once again, much of the task-switching was not the result of outside interference. In fact, sixty-five of the eighty-six switches reported were initiated by the employee him or herself. Most of these switches were related to communications. In the majority of cases, the employee was simply checking to see whether they had received any new messages, even though they had not received any notification of such. All out of sheer habit really—you could even call it an addiction. Just a quick check to see if there's anything new in the inbox.

When something new arrived in the employee's inbox, the employee was usually inclined to respond immediately, especially to messages received on their cell phone. In the case of one particular Xerox firm, the average reaction time was under one minute and forty-four seconds, of which 70 percent were within six seconds. At that speed you barely have time to finish a sentence. We tend to react to incoming messages like we would to the sound of a telephone ringing because we are afraid that the sender might "hang up" if we don't reply straight away. In the example of the Xerox firm, replying to the incoming message also caused employees to open other communication channels or media, as if they were taking a short break from work, one they wouldn't have taken if they had not received the message in the first place. It then took them an average of sixty-eight seconds to return to the work they had been doing originally. This was admittedly quicker than when they had answered a phone call. Nonetheless, after you have been interrupted it takes a little time before you can recall exactly what it is was you were doing and free up your working memory for the task again.

In the case of the Australian firm above where staff spent an average of less than ten minutes working without interruption on a single project, you have to ask the question: is that really enough time to be able to immerse yourself fully in something? Switching regularly between tasks can result in your work becoming superficial because you never get down to the nitty-gritty of whatever it is you are working on. Of course, this depends largely on the kind of work you do (not all jobs require high levels of concentration), but the question is whether we also exhibit the same behavior in situations that do require our undivided attention, and whether we are able to differentiate between the two. Think, for example, of managers who insist that their staff respond immediately to all incoming messages, or companies where employees are required to keep a chat screen open on their computers and who are then praised for their speed and commitment to the cause when their reaction times are quicker than others. You could ask yourself whether these are the kinds of workers who make the all-important, major breakthroughs and contribute most to the firm. After all, we know that many important discoveries have been and are still made in the solitude of an attic, a place that we often associate with the minimum level of distraction and the highest levels of concentration.

A lot of research has been done into the productivity and experiences of workers who regularly switch between tasks in order to find out if there is also a positive side to multitasking. The interesting thing about this research is that it shows that switching between tasks does not necessarily mean that employees spend longer on each individual task, but they do experience more stress and frustration and a heavier workload. This may sound familiar to you, especially if you are prone to working more hastily and becoming easily irritated when you have to put up with constant interruption. My students can tell from how I say the word, "Yes," how often someone has knocked on my office door in the past hour. It is never my intention to act irritated when someone drops by to ask me something, but sometimes it just happens, particularly when I am trying to finish writing an important paper. But, subsequently, I only ever end up trying too hard to make up for lost time, and the result of my haste is invariably a paper of lesser quality.

Regular interruptions can cause workers to adopt a different working strategy and to work in a hasty and less efficient manner. The results of research confirm that task-switching leads to the production of stress hormones like cortisol and adrenaline. Of course, stress hormones can be very useful in life-threatening situations, but they are of little use to you in a normal working situation. It is well known that the production of stress hormones can lead to aggressive behavior, and countless books have been written about the negative effects of stress on our health. Furthermore, constantly switching between tasks doesn't do the atmosphere on the work floor any good either. How do you feel when you have had to put up with interruptions all day at work? The long-term presence of stress hormones in the system often results in crippling fatigue, so don't be surprised if you feel completely shattered after a day of stopping and starting and stopping again.

From the perspective of attention research, large shared office spaces are an odd idea, to put it mildly. Years of research have shown that the brain is very susceptible to distraction. Every time you hear a telephone ringing in the background, your attention will be attracted and your brain distracted. And yet firms still continue to squeeze their employees together into communal offices. That said, it is also true that sharing the same space can lead to improved cooperation and lower accommodation costs, but the question is whether this gain can compensate for the costs associated with the constant distraction of the conversations and movements of your colleagues. After all, each single distraction can cause you to switch tasks because it is sure to draw your attention away from whatever it is you are working on.

Multitasking and Learning

Our brain is designed to learn, even well into old age. Every action that is performed correctly results in a strengthening of the pathways between the neurons involved in carrying out that action. Conversely, pathways that are left unused become weaker over time. This process helps to make our brain more efficient at tasks that are carried out on a regular basis. Remembering information works in the same way too. The relevant pathways are reinforced so that you get better at things like arithmetic, for example. To be able to learn, you have to focus your attention fully on whatever it is you wish to learn. Performing several tasks at once impairs your ability to learn. You can still learn stuff while multitasking, but you will not be able to use the information you have picked up as effectively as you would like to at a later moment, which makes the benefits of learning in this way very short-lived. In fact, you use different parts of the brain when learning during multitasking than when you focus on one single task.

In a study, test subjects were asked to divide various sets of playing cards into two groups. They then had to learn which rules applied to which sets of cards. For some of the sets they were allowed to do so without any distraction, but for other sets the test subjects were required to listen to high- and low-pitched tones on headphones and remember how many times they heard a high tone. Although this distraction did not have a negative effect on the test subjects when it came to learning the rules, it did make it more difficult for them to recall those rules. When they were asked to work with the same sets in a subsequent session, they found it difficult to remember which rule they had learned while listening to the different tones.

Because these experiments were carried out using an MRI scanner, the researchers were able to see which areas of the brain became active when the test subjects were engaged in learning. When they were not subject to any distraction, the hippocampus became active. This part of the brain plays a crucial role in processing, retaining, and retrieving information from the long-term memory. For example, if you want to load the name of an old classmate into your working memory, the hippocampus ensures (hopefully) that the right name is retrieved from your long-term memory. When the test subjects above were learning while listening to the different tones (i.e., while multitasking), the hippocampus was a lot less active or not active at all, meaning that the new information was not retained and could not be retrieved afterwards.

It is worth pointing out that the findings of these kinds of studies are often published in highly respected scientific journals, such as *Proceedings of the National Academy of Sciences*, one of the scientific world's leading publications. The fact that such an important journal decided to publish the findings of the study outlined above has a lot to do with the importance of the concept of learning. If multitasking does in fact have a negative effect on our ability to learn, then this has major consequences for how we educate our children. If the attention of students is constantly switching from one task to another, their education is bound to suffer. Observational studies have shown that students spend very little time concentrating on one single task without interruption when they are studying at home. Even when they were told that they had to study a very important piece of text, they were unable to concentrate for more than three to five minutes on that task. Not surprisingly, social media and chat messages were identified as the main distractors. There was also a strong correlation between the use of social media while studying and performance at school: the heavier the usage, the poorer the performance.

Does Multitasking Make You Dumb?

The idea that multitasking makes you dumb is really just a bit too simplistic, though you might not think so after consulting a much-quoted study recently carried out by Glenn Wilson of the University of London. His study is referenced in many of the books written about multitasking and the effect of new media on our cognitive functions (such as Daniel J. Levitin's influential book, *The Organized Mind*). Wilson's study purports to demonstrate that the mere possibility of multitasking can be enough to lower your IQ by ten points. He also links it to a new phenomenon—"infomania"—and claims that multitasking results in a greater drop in your IQ than when you smoke marijuana. It is a very popular study among the scientific community because almost everyone can imagine what it would be like to suffer a drop of ten points in your IQ.

However, many blogs have devoted a lot of time and effort to questioning the merits of Wilson's study. First of all, it was paid for by Hewlett-Packard and has never been published in a respected scientific journal. The latter is no surprise because the study is built on pretty shaky foundations. For example, not enough test subjects were used (only eight!), and the possibility of multitasking was too often confused with simple distraction. The situation in which the drop in IQ was reported was one that involved a lot of incoming telephone calls and emails. These stimuli cause a person to become distracted and that is not the same thing as generating the potential for multitasking. Everyone knows that an overdose of irrelevant sounds and visual stimuli can have a negative effect on your performance. Even Glenn Wilson was shocked by the amount of media attention his study received. At first, he thought he had landed a nice little earner on the side as a scientist, but, before he knew it, the media and many popular science books were making hay with his ideas. Eventually, he decided to remove the study from his website and replace it with a short document describing how the media had misinterpreted his findings and that he had terminated his research on the subject. Today we can comfortably conclude that your IQ does not drop by ten points just because of the possibility of multitasking.

Multitasking and Studying

Unfortunately, not everything in life is as gripping as your favorite Netflix series. Students find it particularly difficult to concentrate for any length of time when the material they are studying is dull and complex. This is exacerbated by the presence of smartphones and social media in classrooms and lecture halls, as the internet has led to an explosion in the number of potential distractions for the modern-day student. Terry Judd of the University of Melbourne analyzed 3,372 computer sessions by 1,249 students who had been instructed to study on their own without any interference from others. Almost all of the sessions (99 percent) betrayed the telltale signs of multitasking. Although studying took up most of the students' time, Facebook was responsible for a significant 9.2 percent of their time, with 44 percent of all sessions showing log-ins to Facebook. Those students who opened Facebook were also the ones who spent less time working on a single task without interruption and who were most inclined to switch between tasks.

Other studies have revealed that students are able to stick to a single learning task for an average of only six minutes. And there is a good reason for this too. The evidence comes from a study in which a group of students were asked to keep a record of their media usage over the course of twenty-eight days and to report their level of satisfaction at regular intervals. The results showed that students preferred being able to consult various media while they were studying to studying without any access to those media. It also appeared that students who watched television, for example, when studying were more satisfied with how they had utilized their time than students who had not watched any TV. Even though they had achieved less during their study time than they had initially hoped, they still had a good feeling about it. Basically, they found studying to be more enjoyable when the TV was switched on.

The researchers then got thirty-two students to wear a device with which they were to record their activities three times a day over a period of four weeks. They were to use the device to indicate whether or not they were multitasking and, if so, which media they were using. The results of this test also showed that, although the test subjects did not always achieve what they had intended in their study time, the media they used ensured that they felt good about how they had spent that time. Although the presence of various media did result in multitasking, the intention of that multitasking was to satisfy a psychological need.

How should we interpret the correlation between studying effectively and the use of social media? Does the presence of social media really lead to poorer results? Let's take a look at a study carried out in 2007 that examined the effect of chatting on students' concentration levels when doing homework. The scientists expected that, due to the growing popularity of online chatting, the students would be more inclined to multitask and, as a result, less able to concentrate on their studies. In order to establish the effect that chatting had on their study results, the students were asked to fill in questionnaires where they could indicate how much time they spent chatting and how well they thought they were able to concentrate. The results were crystal clear: students who spent more time chatting online while studying found it more difficult to concentrate, and the students who were more inclined to read books during their free time (and who spent less time chatting) had much better concentration.

Based on these correlations, it is very tempting to draw the conclusion that online chatting leads to reduced powers of concentration while studying. The same applies to other studies that have shown a correlation between the number of text messages a student sends during lectures and their final grade, as well as between multitasking in general and the average grades of students. At the end of this book, you will find numerous examples of the above, and there are many more to be found in the relevant literature too. A correlation has been uncovered for every type of media imaginable: Facebook, WhatsApp, Twitter—you name it. And by the way, the most outstanding heading for an article on this topic that I have ever come across has to be "No A 4 U."

The problem with these studies, however, is that they are all based on correlations and usually follow the same method: scientists measure the media usage—using a questionnaire or, for more accurate results, by saving and consulting the computer's history—before measuring the test subject's performance on a certain task and comparing the two sets of results. They then established a negative correlation: the heavier the media usage, the higher the incidence of multitasking and the poorer the performance. However, you cannot conclusively say that one causes the other; correlation says nothing about causality. Although it is tempting to conclude that the student's poor performance is the result of intensive media usage, it is not the only conclusion you can reach. You could also claim the exact opposite: that students will increase their media usage when their grades turn out to be poor—a completely different hypothesis using the very same data.

There is even a third possibility: that there is something else behind the correlation, such as a person's mental aptitude. There may be a very good reason why someone allows themselves to be easily distracted by different media. Some people are less able to concentrate than others and therefore more inclined to be distracted by modern media. One hundred years ago, their counterparts may have done the very same too, but then by nipping out to kick a football around or just sitting and staring into space for a while as a break from their studies. You simply cannot claim that poor grades are caused by heavy media usage. The most you can say is that poor grades and heavy media usage tend to crop up together quite often.

For the very same reason, it is impossible to draw conclusions with regard to food and causality based on the results of a few nutrition studies. There may be a negative correlation between the consumption of a certain kind of oil taken from a certain type of fish and the chances of contracting a specific disease, but that does not mean that eating that fish automatically reduces your chances of contracting the disease. Of course, eating fish is widely regarded as an important part of a healthy lifestyle, just like regular exercise. But the important consideration here is the healthy lifestyle, not the fish oil.

There are ways of correcting for disruptive factors, but in these kinds of analyses it is not always clear where the corrections need to be made. Another interesting detail is that claims regarding causality are almost never explicitly stated in the scientific articles describing such findings, but where they are often found is in the media reports on these studies. To conclude this short lecture on correlations, allow me to share a news item from October 28, 2004, that I found on a Dutch website. It's pretty easy to spot the discrepancy in this item, but my main hope is that you will think of this report the next time you read an article about the benefits of a certain food for your health or the effect that Facebook has on your happiness.

People with No Teeth Have Poor Memory

STOCKHOLM—When a dentist pulls a tooth, it is possible that they are also removing a chunk of their patient's memory, according to the results of a study presented in Stockholm last Friday. "Teeth are apparently very important to our memory," said Jan Bergdahl, a dentist and lecturer in psychology, and one of the authors of the study that followed 1,962 people aged thirty-five to ninety, from 1988 to the present day. The study compared the memory of the participants who still had their own teeth with that of those who had been given false teeth.

"When people have no teeth left, their memory is clearly poorer," according to Bergdahl. The Swedish study does not say anything about the effect the removal of a single tooth has on the memory. The scientists intend to study this further, and also to find out how many teeth a person can have removed before it starts to affect their memory.

How do we go about establishing causality? The only way to do this is by using experimental research techniques in which test subjects are randomly divided into a number of different groups, with each group being exposed to a different experimental condition. The groups have to be made up in such a way that the most important factors, such as age and level of education, are equal for all groups. In this way, you can be sure that any differences in the results between the groups can only be explained on the basis of the experimental manipulation carried out. So if you want to find out whether the use of fish oil has any effect on the likelihood of contracting a certain disease, you will need to create two groups: one that uses the fish oil for a long period of time and one that does not use it (the control group). After the period of time has expired, you establish which test subjects have become ill and compare the percentages for both groups. You also have to ensure that the two groups adhere to similar lifestyles and that the control group does not use the oil, or at least not to any significant extent. Of course, you will understand that, in terms of practicality and cost, a study like this is almost impossible to carry out. After all, it could take decades to achieve reliable results. That said, it is actually the only way to draw absolutely definitive conclusions on the effects a specific substance can have on humans.

For food, a correlational study is probably the maximum that is achievable in terms of feasibility. However, experimental studies have been used to examine things like the effect of media usage on study performance, although up until now these studies have been relatively few in number and pretty basic. An example of such a study is one that was carried out on students' behavior during lectures. The test subjects were invited to attend three different lectures, after which they were asked a number of questions related to the lectures. They were divided into separate groups: a control group that had no access to any media and could therefore devote all of their attention to following the lecture, and a number of experimental groups that were each instructed to use a different type of media during the lectures: email, Facebook, and chatting on their computer or smartphone. All of the experimental groups made more mistakes in the subsequent test than the control group. They had to divide their attention between the lecture and the task of following whatever media they had been assigned and had missed out on some parts of the lecture as a result. A similar study produced the same results, also revealing that the students who had not been distracted had written down 62 percent more information in their notes than the experimental groups, and that their notes were far more detailed as well. If you remember what we said earlier about the usefulness of taking notes at a lecture, you will appreciate why studying without distraction is so much more efficient.

Multitasking is Not Necessarily a Bad Thing

Young people are not the only ones who like to multitask. In a recent extensive study of media usage in the Netherlands, researchers from the University of Amsterdam asked a group of over three thousand Dutch citizens of all ages to keep a diary of their use of various media. There were no major differences in the average amount of time that the respondents spent on multitasking: around a quarter of their day. There were significant differences, however, in the types of media used by the different generations. Young people tended to combine listening to music with online activities (social media or watching videos), and older people combined listening to the radio or watching TV with replying to emails or reading the newspaper.

It is important to point out that not all kinds of multitasking are bad for our concentration. In the first instance, we are not always learning or working on something. And it is not the end of world when you read the newspaper more slowly than usual because you are listening to the radio at the same time. Secondly, the presence of a second source of information does not necessarily mean that it will command your attention. We are more than capable of ignoring a continuous source of information. For example, you can study with the radio on as long as you are able to ignore the information coming from the radio. I remember once having the radio on in my office when a friend of mine came on air unexpectedly. At least that's what I heard afterwards, because it completely passed me by at the time. I had been so engrossed in the paper I was writing that I had paid no attention to the radio at all.

When people wish to concentrate hard on something, they often use a trick that at first seems rather counterintuitive: they put on some music. My parents were never able to figure out how I could do my homework while listening to my favorite Eurodance hits. I have fond memories of listening to DJ Paul, 2 Unlimited, and (my personal favorite) Cappella, while cramming for my exams. And today I still study and write with music playing in the background, although I think (and hope) that my musical tastes have become a bit more refined since then. The next time you stroll into town, take a look in the windows of those hip cafés where students and other young people gather to study or work—many of them will be wearing headphones. In a survey carried out in the Netherlands in 2012, 80 percent of respondents said they listened to music every day at work.

At first, you might think that all that music issuing from all those headphones can lead to nothing but more distraction, to even more stimuli that people then have to try to ignore. However, the majority of workers say that they work better when listening to music. How is that possible? Well, first of all, we cannot maintain a high level of concentration indefinitely. Our concentration tends to falter the longer we try to keep our focus, especially when we are carrying out either very complicated or very boring work. Listening to music can provide the brain with new impulses and so help to keep it alert. Even surgeons listen to music while working—as many as eight out of every ten, according to a recent British survey. Of course, it also depends on what kind of music is playing. In an interview with the Dutch newspaper *NRC Handelsblad*, a surgeon revealed that he never listened to classical music during an operation because he would be inclined to *really* listen to it.

And therein lies a crucial factor when it comes to explaining why listening to music can be a help when we are working: you are not really listening to the music at all. If the surgeon were to listen to his music intently, it would demand some of his attention and require the use of the working memory. He would then be multitasking, and we all know the consequences of that. This influences the kind of music you choose to listen to. If, while at work, I want to listen to a new album by my favorite band, another task is going to suffer as a result. It almost goes without saying that you will be able to work a lot more efficiently when you listen to music with which you are already very familiar or to a playlist with mellow music than when you listen to new music or music that demands a lot of your attention.

Music doesn't just cause more "arousal" (a concept we will be addressing in the following chapter), it also ensures that you are less easily distracted by unexpected sounds. Think back to the students in the hip café, a place where the sounds are wide and varied: conversations beginning and ending, people walking in and out the door, the coffee machine grinding and pouring. By wearing headphones, you can block out all these stimuli and listen only to the ones that you have chosen to let in because they are less likely to distract you. And when your concentration begins to falter, you can just stop working for a moment or two and enjoy the music, which helps you to build up your concentration again. This explains why workers in large, shared office spaces sometimes like to work with earphones in. Of course, this is also because it is not an option to play music loudly on the stereo for the whole office to hear, as that would only annoy their colleagues. Lastly, one's taste in music is a very personal matter. Sometimes I even find myself yearning unexpectedly for a quick dose of Cappella and I have to give in. Bliss.

After reading this chapter, you would be forgiven for thinking that evolution will eventually ensure that we all turn into super-multitaskers. Indeed, there may already be some people out there who don't suffer from any of the problems normally associated with multitasking. If so, are they the harbingers of the changes that the human brain may be undergoing in modern-day society? There are signs that this is in fact the case, but much more research still needs to be done. One study revealed that 2 percent of its 200 test subjects were able to drive in a vehicle simulator without any problem while simultaneously being required to solve mathematical equations and remember a list of words. Ninety-eight percent of the test subjects made major errors in one of the two tasks but the 2 percent had no trouble with those tasks at all. Of course, this percentage is so small that one cannot say anything conclusive about the aptitude of these particular test subjects. They may just have had a really good day, one of those days when everything seems to go right. Or it may be that they have a natural talent for these specific tasks and were able to carry them out automatically (they were not tested on their aptitude for other tasks). When a person is able to carry out a task automatically, they do not need to pay it any real attention and can instead focus their attention on other tasks. This needs to be subjected to a lot more study, but it is interesting to consider that how our society is evolving may be resulting in changes in our capacity for multitasking. That said, evolution is a very slow process, so it will be very difficult, impossible even, to ever prove this.

It would also be tempting to take the findings in this chapter out of context and to predict a gloomy future for our society. But there is no evidence that we have become dumber as a result of all the distractions of modern life. On the other hand, multitasking is certainly a problem when you want to be able to concentrate for longer periods of time. The rise of social media is presenting us with ever more opportunities for multitasking, and we are probably multitasking a lot more than we used to. We still haven't acquired enough hard data, however, and until then we will have to make do with what we know and also keep in mind the limitations of the human brain. The next time you read a dramatic headline about the rise of multitasking, remember that, yes, we are more prone to it than ever before, but that does not mean it is always a problem. Increasing our knowledge about how concentration works can help us to choose wisely when we need to perform several tasks at once. If nothing else, it may help prevent Brian Cullinan from making the same mistake again at next year's Oscars.

3
The Sender: How Do You Hold Someone's Attention?

Picture this: you are poised on the ice, waiting for the starter's gun. The crowd in the arena holds its breath. Hundreds of hours of training, all your diets and nutritional plans, all your mental preparations have boiled down to a single moment, a single ambition: Olympic gold. The starter shouts, "Ready!" A silent pause follows, and then, finally, the crack of the gun. It's time.

The starting procedure described above is used not only in speed skating but in many other sports as well, like track and field and swimming. This is a little strange, really, when you consider that many scientific publications have shown just how unfair this kind of starting procedure is. In fact, it now appears that the result of a race is more or less determined before the skater has even set off on his first lap. To understand this fully, we need to explain the concept of *arousal*. Arousal refers to the level of activation of the central and autonomic nervous system. Or, in other words, how alert you are. When you are sleepy, you have a low arousal, and if you are feeling anxious, your body attains a high level of arousal. The level of arousal influences your reaction time; the higher the arousal, the faster you react. This is what makes the starting procedure so unfair. Speed skaters race each other in pairs and the times of all the competitors are compared at the end. This means that they do not all start at the same time. Speed skating uses a different procedure for starting than the 100 m sprint, for example, and the starter enjoys a certain amount of freedom in deciding how long he will wait—usually between three and a half and five seconds—before pulling the trigger for each race. However, the skater has no idea how long the starter is going to wait after saying, "ready," and therein lies the problem: our body is not able to maintain a high level of arousal for very long. Immediately after they have heard the word "ready," the skater's body is primed to go. However, the longer the skater has to wait until the starting gun is fired, the more their level of arousal will fall. Experiments in the lab have shown that longer waiting times lead to slower reaction times because we react a lot quicker when the body is in a high state of arousal.

I decided to dig deeper into this when the former Dutch Olympic skater Beorn Nijenhuis came up to me after a talk I had given on arousal and attention and told me about the similarities between what I had been saying about arousal and his own theories on the starting procedure for speed-skating. He felt that slow starters (the ones firing the gun) had always put him at a disadvantage but he had never been able to prove it. However, my lecture had just offered him a possible explanation. I was skeptical at first because I figured that the effect of other factors, such as temperature and the quality of the ice, was far greater than the tiny effect that something like arousal could ever have. But we decided to team up with a colleague of mine, Edwin Dalmaijer, and give it some further thought. It turned out to be very painstaking work. We began by analyzing the footage of the men's and women's 500 m speed skating races at the Winter Olympics in Vancouver in 2010. We used the audio tracks from the TV broadcasts to determine, to the nearest millisecond, the amount of time that elapsed between the starter's "Ready!" and the sound of the starting gun. We discovered that the longer the gap between the two signals, the slower the skater's finishing time turned out to be, for both men and women. So, the slower the starter, the slower the skater.

I should point out that we are not talking about the difference between starting times here—not the time between the starting gun and the moment when the skater sets off, but rather the skaters' actual finishing times. After all, the difference between gold and silver is often only a matter of a few hundredths of a second, or even less, as was the case in Sochi in 2014 when only 0.01 seconds split the top two skaters after two 500 m races. In Vancouver, a difference of 1 second in starting times resulted in an extra 0.17 seconds being added to the skater's finishing time. The effect is miniscule, but it is large enough to make a difference in a sport where every millisecond counts. If, in your race, the gap between "Ready!" and the starting gun happens to be a little bit shorter than in the next race, you will enjoy an automatic advantage, however tiny it may be.

The drop in arousal that a skater experiences during a starting procedure is a specific example of something that happens to us all on a regular basis. It is impossible for us to be fully alert every minute of the day—imagine how tense that would make us. In the previous chapters, we saw that multitasking is not a very efficient way of working, but this also begs the question, how long you can stay working on a single task before you start making mistakes or your reaction times start to become slower? This question has played a very important role in the history of experimental psychology. The development of my field of interest went into overdrive during the Second World War when the military became more and more interested in how its soldiers function. They wanted to find out, for example, how long a radar operator could perform his work to the best of his ability and how long a pilot could fly before he started making mistakes. Much of what we will discuss in this chapter has its origins in this period of history.

How Long Can You Keep Your Concentration?

There are many professions in which workers are required to keep an eye on things—professions like customs official, quality controller, or lifeguard. And let's not forget the job of bridge tender. Opening and closing bridges is a pretty routine task involving procedures that follow a standard protocol. However, sometimes things can go badly wrong. In 2008, a fifty-seven-year-old bridge tender

appeared in court in the Netherlands charged with causing a fatal accident at the Ketelbrug bridge in the province
of Flevoland in which a sixty-four-year-old motorist died
when her car toppled into the water below after the draw-
bridge had been raised. According to the public prosecutor,
the bridge tender had not been paying attention and was
guilty of making mistakes at critical moments. The motorist had already driven past the first traffic barrier when she
realized that the barrier at the other end of the bridge had
closed off the exit. She started to reverse her car in panic.
In the meantime, however, the bridge had opened behind
her and the car fell into the water through the gap between
the two sections. The bridge tender stated that he had not
seen the car between the barriers, nor that it had tried to
reverse off the bridge. He was busy at the time trying to contact an approaching boat and was having difficulty doing
so, which meant that his attention was diverted for longer
than usual. After a detailed reconstruction of the events on
the bridge, the court concluded that the accident was not
the fault of the bridge tender. A combination of the design
of the bridge, the color of the car, and the bridge tender's
efforts to establish contact with the skipper of the boat rendered him blameless. The judge declared that he had "not
been excessively careless or negligent" and acquitted him of
all charges.

It is a well-known fact that the risk of an accident like this
one happening increases the longer the employee in question has been working. It is, therefore, very important to
know just how long a bridge tender can maintain a good

level of concentration in order to minimize the risk of this kind of accident. In 1948, the cognitive scientist Norman Mackworth published an article in which he explained how the possibility of a radar operator not spotting an object on the radar is directly related to the length of their shift. To demonstrate this, he carried out an experiment in which test subjects were asked to watch a clock with no numbers for two hours straight. A small circle ran around the clock instead at a constant speed, approximately once every second. Sometimes the circle made a little jump to which the test subjects were instructed to react. After thirty minutes, the test subjects began to make mistakes and the number of mistakes grew steadily the longer the experiment went on.

Regardless of how exciting or challenging an activity is, there is a limit to the amount of time we can fully concentrate on it and give it our undivided attention, and it depends largely on the difficulty of that task or activity. In Mackworth's experiment, the jumps were not easy to spot. So you can imagine that if these jumps had been easier to spot because they were accompanied by a loud sound, the test subjects would not have had any problems successfully completing the two-hour-long task. The point at which your level of alertness becomes a problem depends on the signal that needs to be picked up: the more difficult the task, the more important it is to stay alert.

In Mackworth's experiment, the test subject's level of arousal dropped gradually. Now I don't know whether you have ever taken part in a psychological experiment, but when an experiment is boring, like Mackworth's clock, you can actually feel yourself becoming less alert, slowly but surely. As part of the research for my PhD, I spent a lot of time down in a cellar getting test subjects to carry out very boring eye movement experiments. On the eye-tracker monitor, I was able to see that the test subjects' pupils became less and less dilated and their eyelids began to droop the longer the experiment went on. When it was time to take a break, I would knock on the partitioning wall to wake up the test subject and then engage in a quick chat. That was usually enough to make them fully alert again.

There is a strong correlation between your level of arousal and your performance, something with which you will probably be familiar if you take part in competitive sports or have ever had to give a presentation. If your level of arousal is low, you will not feel the kind of tension that allows you to perform at your peak. You need to feel a certain amount of tension in order to excel. But there is a limit to how much of that you can take. Too much arousal results in stress, and this has a negative effect on your performance. This relationship was first identified by Robert Yerkes and John Dodson and subsequently set in stone in a law that bears their name: the Yerkes–Dodson law. It is a very important law because it tells us which level of arousal will ensure optimal performance during a period of concentration.

The law is based on experiments in which rats had to find their way out of a maze that had only one possible route of escape. When the rats took the wrong route, they were given an electric shock. The idea was to establish which level of punishment would cause the rats to learn quickest. The more

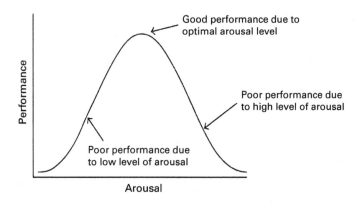

Figure 3.1

The Yerkes–Dodson law on the relationship between level of arousal and performance

the voltage was increased, the quicker the rats learned. But there was a limit to how much they could take, and past a certain voltage the rats' rate of learning began to slow down again until they eventually stopped moving and went completely stiff for fear of receiving more shocks. They even forgot the "safe" places where they were not given any shocks.

Too little arousal is not good for your performance, but neither is too much. In the case of the latter, all of your attention becomes absorbed by the stress you feel until there is none left to carry out the intended task. In many psychology books, this relationship is illustrated using figure 3.1, but the original Yerkes–Dodson law is a little subtler. It differentiates between simple and difficult tasks, as the relationship applies only to difficult tasks. In fact, the higher your level of arousal, the better your performance will be on a task that does not require much in the way of cognitive skills. Yerkes and Dodson differentiated between simple and difficult tasks without actually describing what makes a task simple or difficult. However, recent research has shown that difficult tasks are the ones that demand the attention of our prefrontal cortex, and that an excessive level of stress has a negative effect on the performance of those tasks, especially the ones that make strenuous demands on our working memory.

If you want to optimize the performance of your employees, it is important to motivate them in order to increase their level of arousal, but also to keep their stress levels as low as possible. Of course, everyone has their own critical point when it comes to stress and optimal performance—some of us perform better when faced with a looming deadline, but others don't. There is a point at which our performance inevitably begins to suffer (and this also applies to those of us who always leave things to the last minute). Too much stress is a problem for everyone. I still feel a little tense before my lectures, but I have learned that it helps me to perform to the best of my ability. When I first started out as a lecturer, however, I sometimes "choked" because there was too much tension in my body, with the result that I would clam up and lose my train of thought. This is exactly why tennis players sometimes freeze when they reach match point and why some soccer players find it difficult when they have to take a penalty in an important match. And most of us have probably experienced the other side of the Yerkes–Dodson law as well; trying to perform a task in too nonchalant a manner doesn't work either. Likewise, studying without the required level of alertness will affect your ability to soak up the information.

Do Goldfish Have Better Powers of Concentration Than Humans?

The internet is full of stories claiming that we are progressively losing our ability to concentrate as a result of excessive multitasking and the sheer volume of information we take in on a daily basis. One of the most popular quotes concerns the idea that our span of attention has become shorter than that of a goldfish. The quote in question has its origins in a report published by Microsoft Canada in the spring of 2015 that contained the claim that the human span of attention has shortened dramatically over the past few years owing to the emergence of new media. Whereas in 2010 our span of attention was thirteen seconds, according to the report, in 2013 it had fallen to eight seconds, which is exactly one second shorter than the span of attention of a goldfish. The news was sensational, and it even made the pages of *Time*, the *Guardian*, the *New York Times*, and countless marketing and education blogs. The theory has persisted ever since in the media and is even stated as a fact in the entry for "attention span" on Wikipedia.

The claim is spectacular, to say the least, and may even be considered quite probable by many of us. But is it true? Unfortunately, the study in question can no longer be found online, so I cannot reference it directly in this book. However, there is also a very good reason why it has disappeared from cyberspace. The claim is quite simply ludicrous and has no basis whatsoever in scientific fact. For instance, it turns out that the study did not carry out its own measurements of the human span of attention. Instead, in order to back up its claims, it refers to an unknown website by the name of "Statistic Brain," which, in turn, says that its data come from the National Center for Biotechnology Information at the US National Library of Medicine—a claim the center categorically denies.

In the end, the entire story turned out to be no more than a clever marketing ploy by Microsoft to show advertisers just how adept the company is at attracting the attention of consumers. It is a pity, therefore, that the media were so quick to swallow it hook, line, and sinker. Consequently, it will probably prove very difficult to dispel this myth from the public imagination, just like the myth that we only ever use 10 percent of our brain or that the left side of the brain does entirely different things than the right. The information regarding the span of attention of a goldfish is utter nonsense, not to mention the fact that the "study" wasn't even related to the poor goldfish's span of attention at all but to its memory. Moreover, the memory of a goldfish is actually not that bad at all. In fact, the goldfish is often used as a model for our own memory system and is quite capable of recalling a specific location for food months after their last nibble. Microsoft's claim can therefore be thrown in the garbage where it belongs.

Attention Span During Class

You need a certain amount of concentration to be able to study effectively, and one of the most difficult challenges a teacher faces these days is how to keep his or her students alert during class. In the Netherlands, college lectures have a duration of ninety minutes, punctuated by a fifteen-minute break, and it can be difficult to keep students focused for that length of time. I often show my students the graph shown in figure 3.2 around fifteen minutes into my lectures. The horizontal x-axis shows the duration of a lecture and the vertical y-axis shows the students' level of attention during the lecture. As you can see, there is a spike at the start, after which the level of attention gradually starts to fall before rising again at the end because the students know that we are about to take a break. The graph is more my way of introducing a humorous note to the class and waking my students up than anything else. There is no evidence after all that the graph is in any way accurate, even though it is a common sight in many lecture halls these days.

The theory represented in the graph was questioned in a study carried out by the psychologists Karen Wilson and James Korn, who were fascinated with the recurring claim that the concentration of the average student begins to drop after ten to fifteen minutes. They suggested that, although large differences in concentration levels can certainly be found during lectures, there are many other reasons for this apart from the mere passing of time.

Whether we like it or not, we are all senders of information, like when we're in class or giving a presentation. In order to get your information across, you have to be able to get the receiver to maintain their concentration. And in a world full of distraction, that can be a major challenge. It

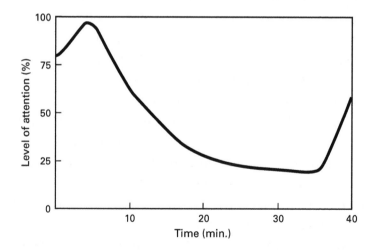

Figure 3.2
The "attention curve" in class

means that the sender has to pay more attention than ever to the content of their message. An interesting story will hold someone's concentration a lot longer than a boring one, and it is the sender's job to ensure that the receiver regards it as a priority that they keep their attention focused on the message they are being given. This is particularly difficult in an age when our smartphones are constantly vibrating with new messages, but I bet we can all name a teacher or TV presenter for whom we will gladly ignore all the buzzing and bleeping just so that we won't miss a single word. The kind of people who have the ability to hold your attention from the first moment until the last.

It is silly to think that our span of attention can be expressed in terms of minutes and seconds. After all, a person's span of attention depends largely on the content of the message being received. Most of us have no problem at all spending hours on end watching a series on Netflix or reading a gripping book without losing our concentration. It all depends on the priority we assign to a task and how interesting or difficult it is. Popular notions like the idea that a You-Tube video should not last more than two minutes are just plain wrong. All you have to do is ensure that your video is interesting enough and that the content matches the interests of the viewer. And there are differences in terms of the platform, too. For example, the average viewing time for videos on YouTube is 870 seconds, and for Facebook it is only 81 seconds. This is because they each have their own specific target audience. We tend to use Facebook when we are looking for quick information from or about friends in our social network or a brief moment of amusement, whereas YouTube is often our preferred media when we want to learn more about a certain subject, enjoy the ramblings of a vlogger, or watch a documentary. Short viewing times have nothing to do with a short span of attention; it is simply a matter of choice. The often-heard advice that you should offer information in short bursts has more to do with the platform where it is being offered and the quality of the information than with the receiver's span of attention. The mountain of information from which we can choose is becoming bigger and bigger, and this makes it even more important to set priorities. Long ago, there was only one TV channel to watch, and so the program makers didn't have to put too much effort into making sure their viewers didn't switch off and do something else. Today, we can choose from an endless stream of channels and information, and attention is the mechanism that forces us to make our choices, something we do on a continual basis. In doing so, we set certain priorities, which is one thing we do have in common with our friend the goldfish.

The Importance of Taking Breaks

It is up to the person sending the information to offer it in as interesting a way as possible so that they can retain the attention of the receiver and keep their level of arousal high. This can be maintained for hours in the case of a Netflix series, but not when it comes to difficult study material, regardless of how well the teacher or author has managed to explain it. On the one hand, we know that we cannot concentrate indefinitely on a difficult task, but on the other hand, we are aware that regularly switching between tasks leads to switch costs and poorer study or work results. So, how should we deal with this apparent contradiction?

The solution lies in the ability to choose the right moment to switch, as was nicely demonstrated by an experimental study into the effect of sending text messages while attending a lecture. The study revealed that the students who sent the most messages scored lower on a memory test taken after the lecture (a difference of 10 percent) and that the level of disparity between the scores was influenced by the amount of time that elapsed between a student receiving a message and then replying to it. Even when other groups sent the same number of messages and some students waited longer to reply, their memory of what they had just learned at the lecture was still better than that of the students who replied immediately to their messages. The difference lay in the amount of time the students took to send a reply. The students who waited longer waited for a moment during which it would be possible to send a message without seriously disrupting their concentration on the lecture. Basically, they chose the right moment to send their text message. So it can be a good idea to switch tasks, but then only when your concentration is beginning to decline.

When I went to visit a group of bridge tenders at work, they asked me how they should handle the issue of cell phones in the control room. Should they be banned during working hours so as to reduce the risk of accidents? No, I said, that would only lead to mass protests. My solution was based on the results of the study described above. They seem to bear out the benefits of implementing a "technology break" in which workers or students are allowed to use their cell phone to read and reply to messages after a specific period of concentration. Merely knowing that you will soon have the opportunity to check your messages—and not at the risk of missing a vital part of the lecture or making a mistake on the job—can provide you with the sense of calm you need to be able to concentrate. It allows you to shut yourself off from the digital world and focus all your attention on the physical world around you. An added advantage of a "tech break" is that it can help to refresh you for the rest of the day, a bit like a coffee break does. For many people, there is nothing more enjoyable than having a moment or two to relax and scroll down through their timeline on Twitter or Facebook. Workers who go outside to smoke usually do so with a cigarette in one hand and a cell phone in the other, so maybe you could think of technology breaks as smoking breaks for people who don't smoke.

This is a useful strategy, not only for lectures or on the work floor, but also for when you are studying. All you need to do is leave your cell phone in a different room so that your concentration won't be broken at those moments when you need to be focused. We all know that many of us are incapable of ignoring incoming messages. This strategy allows you to use the potential reward of new information to raise your level of arousal again after a period of concentration.

It would be easy for me to suggest a "perfect" length of time for uninterrupted concentration, like twenty minutes, but I don't want to make the same mistake that countless pseudo-scientific self-help books have made before. It would be too simplistic to do so because there are big differences in how long individuals are able to concentrate, and the amount of concentration required also differs per task. It is up to each person to find out for themselves how long they are able to maintain an optimal level of concentration.

For this strategy to work, it is important that the technology break is not too long. This represents a major challenge because social apps are designed to hold our attention for as long as possible. These apps keep the user's arousal level high by providing a continuous stream of flashing information. So try to keep your tech breaks as short as possible (set an alarm!), like the way coffee breaks usually last no more than fifteen minutes, after which you will feel sufficiently refreshed to carry on. Some studies have shown that even a break that lasts for as little as one minute can have very positive results.

Although I have avoided including any specific training programs in this book, the lessons it contains about attention do have an important function: learning how to raise your level of "metacognition." The students in the study above who waited until the right moment to reply to their messages were very aware of the value of their attention. And that is precisely what metacognition involves—knowing about knowing, knowing about your cognitive skills and capacity. Being aware of the value of your concentration and the consequences of potential distractions enables you to concentrate in a more efficient manner.

Instead of banning new media from the classroom and the workplace, we should focus on learning more about attention so that everyone can become aware of the value of extended periods of concentration, after which individuals can find out for themselves which strategy best suits their own objectives. Whereas, in the past, students used to stare out the classroom window when they were bored, they now have a tool they can use to remain alert. The solution, therefore, to the problem of multitasking among students is not to create a bland environment free of possible distractions but to actually make use of the refreshing effects of task switching. There are already good examples of how a student's cell phone or computer can be used to keep them focused on the task at hand, like sending them questions during a lecture or inviting them to take part in a poll, for example. At the very least, it might help to rouse a bunch of lethargic students from their slumber.

The conclusion that task switching can help you keep you alert is supported by the results of a study carried out at the University of California in which the activities of forty-eight students were monitored for a period of one week. The test subjects were all fitted with sensors that tracked their heart rate, and their activities on the computer were recorded as well. The students who displayed high levels of multitasking during their computer sessions had a higher heart rate than the students who engaged less in multitasking. Although this was only a correlational study, it is still interesting to note that all of the participants had a lower heart rate during the computer sessions in which they only logged on to social media and did not carry out any other extra activities. It appears that relaxation and social media are interrelated and that there are two possible conclusions: either the reward that we experience through social media results in a lower heart rate *or* we like to fill our moments of relaxation with social media. In the case of the bridge tenders, a few minutes spent on social media after a long period of opening and closing bridges could have a very beneficial effect.

The Sender's Job

In order to communicate successfully with each other, the sender and receiver adhere to an unspoken agreement. The sender does his best to communicate his message as clearly and efficiently as possible, and the receiver agrees to give the sender's message his attention. Both parties are responsible for making the agreement work, and, in a world literally teeming with information, this agreement is more important than ever before. The sender must try to retain the receiver's attention by keeping their level of arousal high. This way the receiver will be less inclined to give in to distractions, such as their smartphone. One of the ways in which the sender can do this is by making sure there is enough variety available to the receiver whenever their concentration is in danger of lapsing.

I have learned to vary the ways in which I provide information in my lectures. I know that students find it difficult to listen to the same story for forty-five minutes and also that I face stiff competition from the devices within their immediate reach. The trick is to anticipate and seize the right moment to change tack. For example, after ploughing through some heavy theory, I often play a video that demonstrates what I have just been telling my students. There are also ways in which you can get your students to use their computers to vote on a particular topic or give a short demonstration in which one of the students takes part in a live psychological experiment. This keeps you, the sender, in control of your own switch moments and ensures that the students' level of arousal (and span of attention) remains high.

This method of information transfer is often compared with the way in which a cheeseboard works. After a four-course dinner, most of us will feel pretty full, but many of us also like to add a cheeseboard as a fifth course at the end. The cheeseboard offers a variety of tastes: sweet, soft cheeses are alternated with hard, salty ones. Diners would quickly become satiated if only one type of cheese was served, but when there is lots of variety, they will keep on eating. The same strategy is used in many talk shows on TV: first, a serious interview, followed by something lighter, a few funny YouTube clips, and finally some music. And the interviews are almost always interspersed with archive footage related to the topic. The viewer is given no time to even think about switching channels because the channel itself is doing the switching for him live on screen, and the switches are directed in such a way that the viewer's concentration does not lapse. The same principles apply to all kinds of communication: the written kind, at the theatre, and during a serious conversation (a quick joke in between to alleviate the tension).

For the purposes of successful communication, you could draw inspiration from the way in which children with ADHD use their powers of concentration. They are usually very good at concentrating on a video game, but not so good when it comes to their homework. Scientific studies have endeavored to find out whether children with ADHD are as good at playing different computer games on a Sony Play-Station as children without ADHD. In experiments, even though children with ADHD scored worse in the standard concentration tests than children without ADHD, their performance on the video games was just as good. There is, of course, a big difference between playing a video game and carrying out a standard task aimed at measuring concentration and working memory.

Video games are designed to absorb the player's attention completely. The scores that can be achieved have a highly motivational effect, and the games are geared to suit the world as a child or teenager sees it. Video games are also extremely fast, and the player is not given a moment's rest. The end result is that the player has no time or space to think about anything else, and the working memory does not have to work hard to retain a specific task, as the characteristics of the task itself are enough to ensure that the player will not become distracted.

The same applies to a tablet. When you use a tablet, you place it more or less directly before you, making it almost impossible for you to become distracted by other visual information. The visual cortex of the brain is continuously bombarded with new visual information, but in this case the proximity of the tablet to the face blocks out almost all of the visual field and concentration is a piece of cake. Have you ever tried talking to a child while they are engrossed in a video game or watching a movie on a tablet? Exactly. Impossible. It is no wonder really that children with ADHD are able to concentrate so well when using such devices.

The studies into the concentration of children with ADHD are not only useful sources of inspiration for learning programs for overactive children, but also for the transfer of information in situations where there are lots of external stimuli. The results appear to show that a good way of transferring information is to literally bombard the intended receiver. In other words, if you want to get your message across, make sure that there is absolutely no way in which the receiver can become distracted by other information. You could also use the kinds of motivational tricks that game designers are so fond of: giving instant positive feedback on performance and using multiple levels to motivate the player to reach the next level, with its attendant promise of even more wonderful stuff on the other side. Who knows how many Olympic medals this might produce someday.

4
The Receiver: How Can You Improve Your Concentration?

The American psychologist Burrhus Skinner (1904–1990) was convinced that concentration can be conditioned, just like you can condition a dog to start slobbering at the sound of a bell. Skinner's work followed in the footsteps of Pavlov, and he was one of the founding fathers of behaviorism. He is most famous for inventing the "Skinner box," an instrument used to study operant conditioning. An animal placed in a Skinner box receives an automatic reward, usually in the form of food, when it executes a certain task correctly. This allows us to study how an animal reacts to a reward and how it learns to associate a certain activity with the potential of a reward.

As well as an inventor and psychologist, Skinner was also a very productive author, publishing twenty scientific books and countless articles during his lifetime. To be able to work hard, he used techniques from behaviorism, such as operant conditioning. Every morning he would begin his daily writing session by ringing a bell in order to create an association between the sound of the bell and his concentration. It was as if Skinner wanted to teach his brain to concentrate all by itself upon hearing the tinkle of a bell.

In actual fact, Skinner engaged in a whole series of rituals early in the morning before his writing session began. He rose at the same time every day and his breakfast always consisted of a bowl of cornflakes. Every morning he read the newspaper and a few pages of the dictionary. Then, at a fixed time, he would go to his study to start his day's work. He always kept the same books close at hand and the same lamp on his desk. Switching the lamp on also set a clock in motion that tracked the time Skinner spent working. And, as if that wasn't enough, he also kept a record of his own productivity. When the clock struck a certain hour, he counted the number of words he had written, just like the Skinner box records the behavior of test animals.

It goes without saying that Skinner liked to measure things, and he is widely regarded as having contributed enormously to making the study of behavior more scientific. The bell was his constant companion, even towards the end his career when it would ring to summon him in the evening to his desk for another hour of writing, and it continued to ring long after he had officially retired. Skinner never stopped working, in fact, seven days a week all year round, and he never went on vacation. He died at the age of eighty-six in 1990 having published his last book only the year before.

Skinner adopted the same attitude towards his concentration as he did towards his experiments. It was almost as if he spent his entire life inside his very own Skinner box. Always that bell, always that amazing productivity. Many people have their own rituals before engaging upon an extended period of concentration. The famous Dutch author Harry Mulisch began every morning the exact same way: a cup of coffee with a teaspoon of cream and a teaspoon of sugar, and a cracker with marmalade on the side. Mulisch was always immaculately dressed, even when he didn't have to venture outside his front door. He stopped working to have lunch—always a boiled egg—at the exact same time every day, and finished his working day by taking a walk in the afternoon following the same route day in and day out. Mulisch was extremely organized in his work, which is not something we can say about everyone. The pianist Frédéric Chopin, for example, used to lock himself in his room for days on end, sobbing furiously while he broke one quill after another in frustration, all in the cause of analyzing and perfecting each note he wrote. It is said that it took him six weeks to complete a single page of music to his satisfaction.

These are but a few examples of the kinds of daily rituals that extraordinary people often follow in their lives. If you read the biographies of influential authors, painters, and composers, you will be sure to find many more. You could view it as a form of procrastination, of course, all that sharpening of pencils and ringing of bells before sitting down to work, but sometimes a ritual is exactly what you need to get yourself fully into concentration mode. In fact, we all have a bell that our brain associates with concentration in one way or another. But although there is no shortage of stories about moments of sheer genius striking people in the middle of the night, most brilliant books and paintings are the result of long, hard hours of work and lengthy periods of intense concentration.

Why Concentration Costs So Much Energy

Why are rituals so important to concentration? You could compare the rituals that precede a period of concentration to an athlete's preparations before a major race. It is important to empty your mind and ignore all potential distractions. Your mental battery must be fully charged. Just like winning a race, a protracted period of concentration requires an enormous amount of effort, but then from the brain. To understand this better, we need to establish why concentration costs so much energy. Today we know a lot more about the processes in the brain that are related to concentration and, strangely enough, most of that knowledge comes from research into the processes that are active when we are not concentrating.

The majority of experiments using brain scanners, such as MRI and PET scanners, examine the activation patterns in the brain while performing a certain task. This enables identification of the brain cells that are used to carry out that task. One of the most important scientific breakthroughs of the past few decades was the discovery of the "default network," a network of regions in the brain that are active during periods of rest. This is a brain state in which you are not actively performing any task; in other words, when your working memory is empty. Up to now, we have mainly been discussing the never-ending battle for and limited capacity of our attention. However, we spend a large portion of our day doing things that do not make any demands on our attention. Taking a shower and cycling your usual route to work are examples of activities in which your default network is active.

It might be a good idea to pause for a moment and examine the term "network." You have probably always been used to thinking of the brain as an organ made up of several different regions, all of which have their own individual functions; for example, executive functions are the domain of the frontal lobe, and basic visual perception is carried out in the occipital lobe. These days, however, neuroscience tends to think of the brain more in terms of networks than in regions. Some functions are not localized in one specific part of the brain but are executed on the basis of the activation of a network of different parts that are all interconnected. Think of the road network in the country where you live. Each individual city has its own special characteristics, but the ability of the country to function as a whole is largely dependent on the speed at which traffic can move from one city to another. The highways connecting the cities are like the pathways in the brain that enable communication between the different regions. (My apologies for excluding towns and villages here. You are all very important, of course, but I don't want to make things any more complicated than they already are!)

As a result of this knowledge, certain neural disorders are no longer regarded as affecting only one part of the brain (unlike some linguistic disorders, e.g., problems with pronunciation, which can be traced back to a specific location in the brain) but more as problems affecting the communication within a network. Communication problems arise when certain pathways die off, which means that the condition of the neural highways starts to deteriorate and the speed at which information can travel becomes diminished. The complications that result from neurological diseases like Alzheimer's are not caused by localized problems but by problems in the communication between specific neural networks. In the case of Alzheimer's, the way in which communication is disrupted is different than for other forms of dementia. In the future, relatively simple measurements of a patient's brain waves may be able to identify which type of dementia they are suffering from, as they can tell us a lot about the state of communication between the various parts of the brain.

Nowadays, test subjects are often required to do nothing more strenuous than lie in a brain scanner, as it is through *resting-state* measurements that scientists are able to see which regions of the brain communicate with each other while at rest. When a test subject is not performing an activity, it does not mean that the brain is inactive. The brain can show spontaneous activation patterns among groups of neurons, and by correlating these patterns with each other it is possible to identify the different networks and see which brain regions are connected (and how strong those connections are). For example, we have long known that the regions that are responsible for controlling our movements are located in both halves of the brain, but we have since discovered that these regions are strongly connected to each other via pathways that traverse the two halves because we have seen that they exhibit the same type of spontaneous activity during periods of rest.

In 1929, Hans Berger, the inventor of the EEG, a monitoring method for recording electrical activity in the brain, revealed that his studies showed that the brain waves of test subjects did not stop when they were no longer performing any activities. His ideas were not taken seriously and for years after it was believed that the brain, or at least a part of it, was only active whenever a task was being carried out. It was only very recently, as late as 2005, that Berger was proven to have been right all along. In that year, Marcus Raichle of the Washington University School of Medicine in St. Louis discovered that the brain only uses 5 percent more energy when performing an activity compared to the situation when it is performing none.

His discovery of the default network was actually a stroke of luck, arising as it did out of another experiment that had gone wrong. In standard neural measurements, the activity when carrying out a task (e.g., identifying whether a letter is a vowel or a consonant) is compared with the activity in a control situation (e.g., where the test subject only has to look at the letter) so that the neural activity during the execution of the task can be "subtracted" from the activity in the control situation. This leaves you with those regions of the brain that were involved in carrying out the task. The control situation must be as similar as possible to the experimental situation so that only the activity related to the task is left over after you have done the subtraction.

In the case of very simple experiments, however, Raichle found it difficult to think of a suitable control situation. So, for want of a better idea, he just took measurements for situations in which the test subject did not have to do anything at all. When he subsequently subtracted the two situations from each other, he saw, to his utter amazement, that the network was not more but actually less active when it was performing the task. After repeating the measurements several times and still coming up with the same result (namely, that certain parts of the brain, such as the posterior cingulate cortex, exhibited the same drop in activity), he knew he had stumbled upon something huge: a network that actually shows a drop in the level of activity when the brain becomes active. Later on, it turned out that of all the networks that are visible during neural measurements, this network is the one that is always there (hence the term "default"). The default network covers a number of the frontal and parietal regions in the brain, regions that are far removed from one another, anatomically speaking. The discovery sparked a revolution. Up until 2007 only twelve scientific articles had ever been published in which the term "default network" had appeared. Between 2007 and 2014 this number rose to 1,384.

When concentration levels drop, the default network becomes active again. The brain cannot maintain both states simultaneously. It has to be one or the other: either the brain is at rest and the default network is active *or* the brain is in an active state and able to perform a task. And this is why concentration costs so much energy; not only does the brain have to ignore stimuli from the outside world, it also has to suppress the activity in the default network. There is a correlation between your performance on a certain task and the activity of the default network—the more the default network is suppressed, the better your performance will be. But what is the point of the default network? If it takes so much effort to suppress it, then this network must surely be good for something, right?

How Daydreaming Can Be Good for Your Concentration

Hopefully it hasn't happened to you while reading this book, but you are probably familiar with the phenomenon of suddenly catching yourself daydreaming while reading. Your eyes travel back and forth across the text, but the information is not being processed. Instead, you are thinking about the vacation you have planned or the argument that you thankfully managed to resolve yesterday. Before you know it, you have reached the bottom of the page, but you have no idea what you have just read. Daydreaming doesn't usually start at the beginning of a chapter but more likely halfway through (maybe even when you started this paragraph), when you begin to get tired. But don't worry, you are not alone.

Daydreaming is taken very seriously within scientific circles, where it is more accurately referred to as *mind wandering*. The level of interest in this area runs more or less parallel to that of the default network, and that is no coincidence either. The neural activity that can be observed when a person is daydreaming is very similar to that found in the default network. The control situation when taking neural measurements is also one in which the brain is not performing any tasks, and so we start daydreaming. We let our thoughts run free and start associating different memories with each other. The American philosopher and psychologist William James once famously described our thoughts as a stream of consciousness that flows in every direction, a stream that is sometimes very strong and at other times weak. When we are daydreaming, memories that we thought were lost forever can come to the surface again, or you may suddenly find yourself realizing that you have forgotten someone's birthday—exactly the kinds of things that don't happen when you are concentrating hard on your work. And just like the default network cannot be active while we are performing a task, neither are we able to daydream and do something complicated at the same time.

Daydreaming is always there waiting to pounce on us when we least expect it. A boring lecture or meeting, or a loss of attention due to fatigue can send you spinning off into daydream land. When I am giving a lecture, I sometimes find it difficult to tell which students are daydreaming and which are not, apart from the ones who are staring out the window, of course. There are a number of functions that are associated with daydreaming: making plans for the future, feeling empathy, and reflecting on your role in a particular social context. Daydreaming is thinking about yourself in a carefree manner or considering the long-term consequences of an incident and then suddenly realizing that you have missed an appointment.

We actually spend a lot of our waking hours daydreaming. Scientists at Harvard have developed an app that can be used to ask test subjects what they are doing at any given moment during the day. The technique is called *experience sampling*. In experiments, test subjects were also asked whether they felt happy at that particular moment. The scientists are still collecting additional data, but a whopping 2,250 people took part in the first study, so they have already compiled a massive amount—approximately 250,000 measurements. (By the way, you can download the app at http://www .trackyourhappiness.org.)

The study showed that the participants spent 47 percent of their day daydreaming instead of working on the task they were supposed to be carrying out. When the test subjects indicated that they were thinking of something other than their intended task, they were asked whether they were thinking happy, neutral, or unpleasant thoughts. Later on, when the article was published in *Science* magazine, the headline, surprisingly enough, was, "A Wandering Mind Is an Unhappy Mind." The participants indicated that they were a lot unhappier when they were daydreaming than when they were performing a certain activity. Their thoughts while daydreaming were not, for the most part, the happy kinds of thoughts that we would normally associate with daydreaming. It appears that although you are likely to have happy thoughts when walking on a beach and watching the sun go down, this tends to be the exception rather than the rule.

These results have interesting consequences for people who, in their pursuit of the perfect life, believe that never having to work a single day again would make them very happy. Although this is very much open to interpretation (happiness is a relative concept and one that is difficult to measure, and the sample taken was not entirely representative of the population as a whole), it is quite possible that we are a lot happier when we are fully immersed in our work than when we are daydreaming. This idea can also be found in many forms of relaxation therapy in which people are advised to focus on the activity currently occupying their mind (so that you can "be in the moment"), and the above study is often cited in support of such techniques. As the scientists themselves so succinctly concluded, "A human mind is a wandering mind, and a wandering mind is an unhappy mind." An additional and unexpected discovery during the study was that test subjects reported feeling very happy when making love and that they almost never daydreamed when doing so, which makes this study all the more remarkable, given that the test subjects would have had to pass this information on via their smartphone while they were busy between the sheets—all in the name of science, of course.

There are significant differences between individuals when it comes to how much time we spend daydreaming. We are all familiar with the image of a child staring out of a classroom window into the distance. You may even be a dreamer yourself. Based on what you now know about the benefits of concentration, you will not be surprised to hear that daydreaming can have a negative effect on your ability to perform a specific task. However, it may come as a surprise that, on average, people who are inclined to daydream a lot have a lower working memory capacity and score worse in IQ tests. That said, we must remember that we are talking about a correlation here and it does not necessarily mean that daydreaming leads to a lower level of intelligence, although there is a strong relationship between the two. And that is not surprising—you need a good working memory to be able to maintain your concentration, after all. Given the fact that you suppress your default network when you are concentrating, this automatically means that your default network becomes activated again when you suffer a lapse in concentration, and then you start daydreaming. You could say that daydreaming is actually concentration gone wrong, and daydreaming can make you less aware of your surroundings too, which can be very dangerous when you are driving a car. This diminished awareness of your surroundings is known as *perceptual decoupling*, meaning that your sensory perception gets cut off from the outside world.

In another excellent daydreaming experiment, test subjects were asked at random moments while performing a task whether their attention was actually focused on the task or whether their mind was somewhere else entirely. Not only did the results show that the periods of reduced attention were also the periods during which the most mistakes were made, but they also offered a possible route out of the daydreaming maze: the prospect of a reward. When test subjects knew that they would be given a reward if their performance was good, not only did their performance improve, but they were also less inclined to daydream. So the next time you want to concentrate intensively on your work and avoid daydreaming, you could try promising yourself a reward.

So is it all bad news when it comes to daydreaming? Well, some scientists believe that daydreaming also has some very important functions. For example, when you are less preoccupied with the world around you, it is easier to focus on yourself and make plans for the future. When people are asked what their daydreams are about, many say that they often concern personal matters, otherwise known as *autobiographical planning*. These kinds of thoughts can produce many benefits, especially when the task on which you are supposed to be focusing is not so important or does not require all that much attention.

Another benefit of daydreaming is that it can make an extremely boring task (even more boring than brushing your teeth) more enjoyable. After test subjects had been asked to perform a very tedious task for forty-five minutes, they reported feeling less happy than they did beforehand. However, the drop in the level of happiness was less pronounced among those test subjects who reported having daydreamed during the task. A potential solution for boredom is often included on the list of possible functions of daydreaming. You could regard the brain as a machine that always has to be doing something. So whenever you have nothing to do and are just killing time, you can always allow your thoughts to whisk you away briefly to some imaginary future.

With regard to the aforementioned function of daydreaming, something does seem to have changed in recent times. It appears that since the arrival of the smartphone we have no reason to be bored anymore. Diversion is always within hand's reach. This change has not been subjected to scientific study (basically because that cannot be done, given that you can't go back in time), but it is not far-fetched to conclude that since the arrival of the cell phone we have stopped daydreaming as much as we used to. We receive so many stimuli via our phones that it is actually quite difficult to daydream these days. In fact, tracking social media on your smartphone is a kind of wasteland between concentration and daydreaming: you are not fully concentrated on the task you are carrying out; instead you allow all the information and stimuli to wash over you without having to undertake any real action. On the other hand, you are not daydreaming either, because you still have to process all that information. So, given that daydreaming can also have an important function, it might be a good idea to allow yourself to become bored from time to time and to ignore your smartphone.

One of the most important possible functions often attributed to daydreaming is the stimulation of creativity, thinking up new ideas and taking the time to solve complicated problems—the power of the unconscious. Books about concentration and creativity often advise readers to daydream and let ideas appear of their own accord. The theory goes that when you are daydreaming, your unconscious mind goes about solving your problems for you. In fact, it would be better to leave this kind of thing to your unconscious mind altogether instead of trying to solve your problems with conscious thought. This points to another major problem with social media: we are becoming less creative because we are daydreaming less. But is this really the case? There is no doubt that taking a break during work so that you can come back later and tackle a problem with a clear mind is a good idea. But what about the claim that while you are on your break, a whole bunch of unconscious processes are busy solving the problem for you?

This claim regarding the power of the unconscious mind is based on the findings of the psychologist Ap Dijksterhuis, who has written a number of influential scientific books and articles on this subject. His findings are surprising, given what we already know about the human brain. The functions of calculation and reasoning are the domain of the working memory and they require concentration. The working memory is the place where all of the information in the brain is brought together and where the tools we need to be able to consider that information are located. This is also the information of which we are conscious, and on the basis of these definitions there cannot be any such thing as "unconscious thinking." After all, that would mean that we also have an "unconscious" working memory that is just as powerful as our conscious working memory. In recent years, the findings of Dijksterhuis and his colleagues have come under increasing scrutiny (see below).

The Not-So-Clever Unconscious Mind

Ap Dijksterhuis claims that conscious processes lead to better results in the case of simple choices, but that more complicated choices, like which new car you should buy, are better left to the unconscious mind. This is referred to as the *deliberation-without-attention hypothesis*. If you want to know which car you should buy, just start daydreaming and your brain will figure it out for you; in no time, the answer will come drifting up from your unconscious mind. It goes without saying that we often make irrational choices and don't always pick the best solution. This makes the idea that the unconscious mind might be better at making those decisions a rather attractive proposition. After all, the capacity of the working memory is limited, which makes it hard for us to solve complex calculations in our head, for example. However, it is very doubtful whether the unconscious mind is capable of solving such calculations.

In an experiment, Dijksterhuis asked test subjects to read information for four different cars. Each car was described on the basis of four or twelve features. For one of the cars, the majority of the features were rated as very good (clearly the best choice), and there were other cars that scored a lot worse. In the conscious choice situation, the test subjects were given four minutes to decide which car they would choose. In the unconscious choice situation, the test subjects were instructed to use the four minutes to solve a number of difficult puzzles.

The results were seen to match the deliberation-without-attention hypothesis. When only four of the cars' features were provided, the test subjects in the conscious choice situation picked the best car more often than the test subjects in the unconscious choice situation. The results were the opposite when twelve of the cars' features were provided and the choice was more difficult. In that case, the unconscious choices proved to be the best ones. The results were endorsed by studies of buyers at Ikea (furniture being difficult to choose) and de Bijenkorf, a well-known Dutch department store (clothing being easy to choose). Customers who had carefully considered their purchases at de Bijenkorf were a lot more satisfied with their product a few weeks later than the customers who had purchased their items on impulse. The results were the other way around for Ikea: the customers who had engaged in impulse buying were more satisfied than those who had given careful consideration to their purchases.

My tone of voice may have already betrayed my cynicism towards these experiments, but not only are the findings based on shaky "facts," they are also very difficult to replicate. In 2015, a team of researchers led by Mark Nieuwenstein carried out an extensive study to see if they could come up with same results as Dijksterhuis's crew. The original Dijksterhuis study had used quite a small number of test subjects per experiment, so Nieuwenstein and his colleagues decided to use ten times as many. Apart from that, they used the exact same list of features and the same diversionary tactics. Even though the researchers based their findings on the data collected from 399 test subjects, they did not see any effect that even closely resembled the original findings. They also analyzed all of the other studies that had published evidence either supporting or refuting the theory and came to the conclusion that previous studies that came out in favor of the theory had also used very few test subjects, which made the findings less than reliable.

These studies are plagued by other problems too. For example, the researchers apparently decided for themselves what constitutes a positive feature in a car (e.g., the presence of a cup holder was considered to be equally important as the car's fuel efficiency), they used students to decide which cars were worth buying, and they gave the test subjects in the conscious situation a lot of time to consider their options (four minutes for sixteen features). In a different study that examined this specific aspect, test subjects were already able to make their decision after thirty seconds, a lot less than the four minutes they were given. This shows that the Dijksterhuis's results might be due to the negative consequences of being given too much time to think and not from any ability to make unconscious decisions.

The whole affair is indicative of the replication crisis currently afflicting social psychology in which it has become clear that it is difficult to replicate the findings of earlier studies. I would like to emphasize the point that this does not necessarily mean that previous studies are in any way fraudulent (a charge that is often unfairly made). All it may mean is that there are problems with how a particular study was set up; for example, the number of test subjects. In any event, all any discipline can ever hope is to learn from these kinds of mistakes. That is the best way to move science forward, and social psychology is now one of the frontrunners when it comes to ensuring better methods and statistical analyses. It would also be a great help if people outside the scientific community became more aware that claims that have been published previously by the media and that have played an important role in relation to theories on creativity may, in fact, be incorrect and in need of further corroboration. For what it's worth, here's my contribution: you cannot make better decisions with your unconscious mind than with your conscious one. Regardless of how irrational you are, you have to think consciously about something to be able to arrive at a decision. And that's a fact.

I should also point out that Ap Dijksterhuis continues to defend his theory and insists that each time there has been an unsuccessful replication, it is because the wrong task has been used. The most we can say is that his theory holds only in very specific situations.

In my opinion, there is no evidence to suggest that the brain can solve problems on its own while you are taking a mental siesta. However, restarting a thought process can help you to approach an issue from a fresh perspective. For many of us, a mental pause can have a positive effect, hence the often-heard suggestion that we should "sleep on it" when faced with making an important decision. Switching your attention to something else can give you the time you need to approach a matter from a different perspective and maybe even reach a different conclusion. Sometimes you can become very fixated on one single detail and a mental pause can make you stop, step back, and think again. Maybe you were very taken by the color of the new car you saw yesterday in the showroom, but after a night's sleep you realize that the upholstery doesn't suit your tastes at all. Nothing magical has happened in the intervening period, you are merely benefitting from a fresh perspective. A period of daydreaming can help you to recharge your mental battery, just like the technology break in the previous chapter. The brain can become exhausted from the effort involved in trying to ignore all of the stimuli, both internal and external, with which it is constantly being bombarded. But what should you do during a mental pause? Spend the time on your smartphone or go for a walk?

How to Recharge Your Battery

The composer Ludwig van Beethoven was a man of habit. For breakfast, he always drank a cup of coffee made from exactly sixty coffee beans. He would count them out carefully one by one because he was convinced that sixty was the perfect number for the perfect cup of coffee. He would then spend the day working at his desk and taking long walks. Elsewhere in Europe, the philosopher Voltaire spent every afternoon driving around his estate, and the painter Joan Miró worked almost exclusively in the Spanish seaside town of Mont-roig del Camp.

Is there are a reason why many great philosophers and artists seek out nature in order to concentrate? One possible explanation can be found in the *attention restoration theory*, or ART for short. This theory claims that the environment in which you find yourself determines the extent to which your attention can be regenerated and that nature is the best place for doing so. ART even goes so far as to claim that nature is in itself a form of therapy. One without any side effects and completely free of charge!

So is there any truth to these claims? As we have already seen, there are more or less two kinds of attention: involuntary attention, the kind that is driven by information from the outside world, and voluntary or focused attention, which is driven by your goals and the things that are important to you at any given moment. We also know that voluntary attention determines your ability to concentrate. The idea behind ART is that a walk through natural surroundings can restore your voluntary attention. Nature is full of intrinsically intriguing information, such as sunsets and fabulous views. This information does not require any direct action from your voluntary attention. In addition, your involuntary attention is also taxed to a significantly lower degree than it is in an urban environment where you are constantly being distracted by other people and stuff like advertisements and traffic. In a city, you need to use far more focused attention to find your way around and ignore all of the distractions. In other words, walking around in nature demands less of your voluntary attention than maneuvering your way around town. In nature, you can just take it all in as you stroll around (except for in the jungle, maybe) and allow your attention to be drawn to whatever catches your eye without having to work hard to ignore all the other information.

This explains why going for a walk can be so beneficial to your level of performance, and the improvements can be found in all kinds of activities related to different aspects of attention, even memory. You don't even have to go hiking deep into the mountains to achieve these improvements; just looking at images of nature can produce the same results. And it has nothing to do with nice warm weather either, as the same results can be found in the freezing cold of winter too.

These findings have interesting implications for training programs aimed at improving performance. Imagine you have to perform a task that demands your full attention, as is the case with almost every activity that requires concentration. If you have just experienced great difficulty maneuvering your way through traffic before carrying out the task you are now faced with, your performance will probably be a lot worse than if you had just come back from a long walk in the woods. This is definitely worth remembering the next time you have to take an exam. I always feel that students who come running into the exam hall at the very last minute because their train was delayed have far more difficulty concentrating than the students who arrived refreshed and on time.

ART offers interesting insights into how we could improve our children's learning environment. Children with ADHD find it easier to concentrate after they have spent twenty minutes walking in a park than when they take a twenty-minute walk around town. The improvements revealed by this particular study were comparable with those reported for methylphenidate (i.e., Ritalin), a medicine that is prescribed for many children with ADHD. It is quite likely, therefore, that schools located in natural surroundings are better for children's concentration than schools in urban areas. Even windows that offer a view of nature are good for your concentration. If you also consider the positive effect that nature is known to have on people who suffer from depression, then it is easy to see why Beethoven liked to go for long walks before sitting down to work on his next masterpiece.

You Can Train Your Concentration

I have already discussed how retaining a task in the working memory requires you to ignore all other intrusive information. Think of a bouncer whose job it is to prevent unwelcome guests from entering a nightclub. At a certain point, the bouncer may become tired and the task may disappear from his working memory. The good news (not only for the bouncer but for the rest of us too) is that it is possible to train your concentration. An excellent review article published by the psychologist K. Anders Ericsson cites a number of examples of extreme expertise in the case of complicated tasks, such as playing the violin. According to Ericsson, when you are starting to learn the violin you can only practice effectively for a maximum of one hour each day. You will not learn more by practicing for longer. The more experience you gain, however, the longer you will be able to practice. In the article, he refers to violinists at the Berlin University of the Arts. These gifted musicians practice for a maximum of four hours a day, often in two sessions. The same approach has been found in other areas of expertise too. A special talent for a specific skill is only half the story; athletes and musicians who perform at the highest levels also train extremely hard. The trick is to learn how to maintain a consistently high level of performance for increasingly longer periods of time. If you don't practice, you will never become a good violinist, and this applies almost certainly to all tasks that require concentration. Your capacity to study long and hard can also diminish when you haven't done it for a while. In that case, you will need to undergo a training program to build up your level of expertise again. It's not just about having the ability to perform a certain skill at a high level but also about being able to perform at that high level for increasingly longer periods of time.

It is also true that working endlessly for hours on end rarely produces the desired results. In the case of an extremely difficult skill like playing the violin, there is little point in practicing for more than four hours a day. This might sound reassuring to some, but to others the idea of being able to concentrate for as long as four hours is a distant reality, especially when their working day is full of meetings and other ancillary activities. At my university, I only ever manage to concentrate for four hours or more on my work during the summer when there are fewer students around. And that is why, during the academic year, I often reserve a few hours in my schedule whenever I can and spend them working among students in the peace and quiet of the university library.

Although concentration is certainly a matter of training, you should try to avoid apps that promise to improve your attention through training programs that require you to perform tasks on your smartphone. The downside of these programs is that they do not apply to external situations and are of no use to you when you are not using the app. You might get a kick out of improving your score on the app, but it won't help you much out there in the real world. A good example of this can be found in an extensive experiment in which scientists made use of the large viewing figures for the popular BBC science TV program *Bang Goes the Theory*. No fewer than 52,617 participants between the ages of eighteen and sixty took part in an internet study lasting six weeks. Beforehand, they were tested on a wide range of cognitive functions, including attention. The tests were used to map their cognitive functions and record any improvements. The experimental group was instructed to complete six training tasks in ten minutes three times each week. The better the test subjects began to perform, the more difficult the tasks became. One of the tasks focused specifically on attention, and the others were related to cognitive functions like memory and reason. The control group spent their time at the computer too, but not doing anything that would have any training effect.

After six weeks, 11,430 participants had completed the training and they had their attention measured again. None of the training programs (including specifically for attention) led to any actual improvements in relation to attention. So, although the test subjects did perform better on the training tasks than the control group, this training effect did not generalize to any other attention tasks. The only way to improve your attention in relation to your studies or work, therefore, is simply by studying or working more. And who knows, you might even get some actual work done in the process too.

The Value of Meditation

It is worth our while taking a closer look at the subject of meditation, especially given the number of scientific studies currently claiming that it can help to improve your concentration. The aim of many meditation exercises is to help you to relax and focus your voluntary attention on internal mental processes. It is difficult, however, to provide an exact definition of "meditation" because it is not one single type of training but a collective term for a large number of different techniques. Many modern meditation techniques have their origins in Buddhist traditions that stretch back thousands of years.

There are different styles of meditation that are relevant to concentration, including a kind known as *focused attention meditation*, the aim of which is to focus your voluntary attention on a certain object for a long period of time. That could be an object in the real world, such as a vase, an imagined object, or just your breathing. The idea is to keep a very close eye on your attention and to maintain focus. This trains your concentration. Intrusive information, a thought or a sound, must be ignored, and if you cannot do that, you must try to detach your attention immediately from the "intruder" and focus once again on the main object. The connection with concentration is immediately apparent: if you are working in a shared office space and do not wish to be distracted, you will need to use the exact same skills. Another, more open kind of meditation is one in which you focus your attention not on an object but on everything that you hear, see, or feel without reacting to it. Both kinds of meditation can be used to train your attention.

It is by no means fanciful to suggest that these kinds of meditation exercises are good for your concentration. After all, they have nothing to do with ley lines or other pseudoscientific beliefs but are based purely on the idea that meditation can be used to train your concentration. All you have to do is picture the place where meditation is usually practiced—a peaceful spot with little or nothing to distract you. Only very experienced practitioners are capable of meditating in busy places. There are other benefits attached to meditation too, of course, but I am focusing on how it can benefit your attention. I will ignore everything else for the moment. Very Zen.

It is only in the past fifteen years that serious studies have been carried out to examine the positive effects of this age-old form of training. In these studies, test subjects are sometimes sent to a retreat for several months to participate in intensive meditation training. The training often involves up to ten hours of meditation a day. In other studies, the approach is less intensive, but they still require several hours of meditation each week. It is important, however, to point out an obvious limitation in the effectiveness of these studies: the kind of people who tend to participate in these experiments. It is essential that the test subjects have no previous experience with meditation, but you surely need to have some level of interest in it to agree to take part in such an experiment. As a result, you can never conclude that the positive effects of meditation apply to everyone, because the studies never include those people who have no interest whatsoever in meditation.

There is no denying, however, that research into meditation has delivered positive results in relation to concentration. Although there have not been any studies yet involving a large number of test subjects (simply too expensive), inexperienced test subjects who took part in intensive training subsequently scored better on tasks that measured their concentration, and this applied to both kinds of meditation described above. Think back to Mackworth's boring clock experiment: your performance at such a task is bound to improve after you have spent three months meditating for five hours a day.

This might make you wonder whether it may be just as effective to practice watching Mackworth's clock for three months instead of doing all that meditation. Maybe so, but the problem with many training programs, as I have already mentioned, is that the results remain limited in scope. You can get better at solving a particular puzzle by doing it over and over again, but that does not automatically mean you will be better at solving other puzzles too. The famous taxi drivers of London, with their incredible capacity for memorizing the streets of the city, do not perform better at other memory tasks. In fact, there are very few training programs whose results can also be applied to other activities.

Meditation appears to be an exception to the rule. Test subjects become better at performing tasks that do not require any meditation. Heleen Slagter at the University of Amsterdam has compiled a list of characteristics of meditation that may explain why this form of training appears to work so well. These characteristics are not the exclusive property of meditation, by the way, and it is quite possible that some forms of training that have the same characteristics, such as music courses, produce the same positive effects.

1. The context of the training. There are several different processes attached to meditation. For example, in the case of focused attention meditation, the attention must be focused on an object for a long period of time and all possible distractions must be ignored. This helps to train different elements of attention. Optimal performance in your daily activities often requires the simultaneous use of different aspects of your attention.

2. The variation in the task. In many meditations, attention must be focused on a different object in each separate training. This can be an object in the room, or a thought or feeling. This helps to train different variations of the same task.

3. The type of cognitive process (attention) being trained. As should be clear by now from this book (and my previous one), many of the tasks we perform each day demand our attention. Training our attention, therefore, can be of great benefit to a wide range of activities.

4. The task's level of difficulty. Training sessions are especially effective when the task becomes more difficult as you become better at it. This helps to maintain the element of challenge in the process. In meditation, your teacher gradually guides you toward more complicated forms of meditation.

5. The arousal of the student. As we have already seen, good concentration requires the right level of arousal—not too much and not too little. In meditation, the trick

is not to fall asleep, while at the same time keeping your enthusiasm in check.

6. The duration of the training. The training programs in which test subjects take part can last for months, with long sessions every day. There is almost no other kind of training that students are able to follow for so long.

Meditation would seem to offer a good solution for day-dreamers. In experiments, test subjects were seen to engage less in daydreaming after a long period of meditation training. This appears to show that meditation enables you to suppress your default network for longer stretches of time, thereby improving your ability to concentrate for longer. Research into meditation is still in its infancy, and although there are many obstacles that need to be overcome, none appear to be insurmountable. The research may even help in the development of computer games whose skills can also be used for activities that have nothing to do with the game itself.

Going Offline

One way of keeping your concentration sharp is by taking a break from whatever task you are performing. This can have major consequences for your work situation. A study carried out by a well-known US consulting firm, Boston Consulting Group, revealed that some workers spent up to twenty-five hours of their time outside office hours each week keeping tabs on their email in the belief that it was important to be able to react immediately to questions from clients or colleagues. This same company decided to experiment with mandatory periods of free time during which workers were not permitted to check for new emails or (voicemail) messages. These periods fell both within and outside office hours and were part of the planning at the start of each new project.

The experiment was so innovative that, at the start, workers had to be literally forced to take the time off. They found this particularly difficult when faced with a fast-approaching deadline. In the first experiment, the workers were required to take one full day off each week, preferably in the middle of the week. Given that the employees were now using only 80 percent of their working time, extra team members had to be recruited to keep working hours at their original level. At first there was a lot of resistance, partly because some employees thought that participating in the project would be bad for their career. But the experiment was a great success and the level of enthusiasm led to an even more extensive second experiment.

In the second experiment, a number of teams were required to insert a free evening into their schedule. No extra manpower was recruited to fill the gap. Of course, the employees were also free during the other evenings in the week, but on this particular evening they were strictly forbidden from doing anything that was related to their work. The workers came in to work the next day feeling fresh and raring to go. In questionnaires, they indicated that the communication within their team had improved and that they were more satisfied with the balance between their work and their private lives. They even suggested that there were noticeable improvements in the quality of their end product. So it appears that you don't need to be on call 24/7 in order to keep your clients happy.

Continually checking for new messages only causes you to switch more frequently between tasks outside of work as well, even between tasks that only lead to more work stress. Until you check a message, you cannot know what it contains or who it's from. It could be an emotionally charged message from a colleague, for example, that upsets you at the very moment when you are on the floor building a Lego castle together with your child. The chances of you returning to your Lego duties are pretty slim if you open the message, read its contents, and allow it to irritate you. You are more likely to ignore your child and reach for your laptop instead.

This kind of behavior is typical of workers who feel obliged to make themselves contactable at every moment of the day and night. It is not very surprising, therefore, that on January 1, 2017, a law came into effect in France enshrining the worker's right to go offline outside of office hours. The decision was based on a study commissioned by the French employment minister in 2015. The results warned of the risk of burnout, insomnia, and relationship problems because workers were being increasingly required to make themselves available outside of office hours. The legislation is not exactly watertight, however, as the employee has the responsibility of reaching an agreement with his or her employer. In addition, the employer can only be penalized when the employee has taken them to court and is able to prove that his or her burnout is the result of continuous telephone calls and messages outside of working hours.

The initial reaction to the legislation in France has been one of skepticism, to say the least, and the rest of the world had a little snigger at the expense of the crazy French. Viewed in the light of the research results in this chapter, however, it has to be regarded as a step in the right direction. It may not be much more than a case of political maneuvering, of course, but it is the right kind of maneuver nonetheless. We do not live in times where people take to the streets every day to protest for better working hours and more holidays, and so a new approach is required. Examples of this new approach are offices where the mail server is shut down when the office closes, and even a company in Amsterdam where the desks are literally hoisted into the air at five o'clock to call a halt to the day's work. Just two of the ways in which we are attempting to regain ownership of our work/life balance.

Other Ways of Improving Your Concentration

There is one more factor related to improving your concentration that we have not mentioned yet, but for which the evidence is very convincing: fitness. Research has shown that a physically fit person can concentrate better.

For example, a period of regular aerobics sessions caused test subjects to perform better at attention tasks compared with test subjects who had not taken any physical exercise, and the findings apply to people of all ages. A brain in a fit body is more capable of ignoring intrusive information and suppressing the default network, which makes it easier to concentrate for longer periods of time. So if you want to improve your powers of concentration, regular exercise is highly recommended.

Although being fit is certainly not the only factor on the list of additional ways to improve your concentration, the effectiveness of other methods has yet to be scientifically proven. Positive results have been recorded, however, for various kinds of electrical and magnetic brain stimulation techniques, such as transcranial direct current stimulation (tDCS), and students have been known to use the same kind of medication taken by people with ADHD in order to improve their concentration. The right kind of brain stimulation has yet to be found, however, and although there seem to be plenty of people who react well to the use of methylphenidate, the large individual differences mean that it is not yet a suitable concentration remedy for people who do not have ADHD. Future studies may one day reveal the right methods, but until that time it is wiser to avoid using such methods. Incorrect use can lead to issues like physical harm because of the side effects of the medication, for example. In this case, unfortunately, the old adage, "Nothing ventured, nothing gained," does not apply.

Receivers of information have the difficult task of maintaining their concentration, whether that is during a creative process or when reading a complex book. Concentration requires a certain amount of maintenance work, too, and science has revealed methods that can help: meditation, training, and fitness. Also, during periods of concentration it is important to take regular breaks, preferably in natural and peaceful surroundings. Switching to autopilot for a moment or two can allow you to recharge your attention before you embark upon your next work of art or project.

5
The Importance of Concentration in Traffic

One Friday afternoon, on October 21, 2011, Koen van Tongeren was driving home on the A4 near The Hague. He remembered he needed to call someone about a job and looked up the number in his cell phone. While he was busy searching for the number, a traffic jam started to develop up ahead, but Koen failed to notice because he was still trying to find the telephone number. He slammed full speed into the stationary car in front of him. A mother and her two children were in the car. The mother suffered serious injuries but she survived the accident, as did her eldest child; however, her two-year-old son who was sitting in the back did not. Koen was sentenced to 150 hours of community service for reckless driving and had his driver's license suspended for six months.

Although his sense of grief probably pales in comparison with that of the victims' family, Koen's life will never be the same again. He had a great life, as he described it himself to the Dutch media, a good job, lucky in love, and a wonderful home. Now, however, he will have to shoulder the blame for the death of a young child for the rest of his life, all because of a few careless seconds in which he failed to concentrate on the road. Since the accident, and with the permission of the victim's family, he has spent much of his time sharing the story of the accident with others, including in several documentaries. He also features in an advertising campaign for an insurance company that shows a number of drivers being asked to perform a task on their smartphone while driving a car on a test track. While they are doing so, a cardboard car appears on the track and stops abruptly. After the drivers have crashed into the cardboard car, Koen appears on screen and tells his tragic story. A story of what can happen when you take your eyes off the road, even just for a split second.

Whereas a car offers you some kind of protection in traffic, on a bicycle you are much more vulnerable. Tommy-Boy Kulkens from the town of Kortenhoef was a self-confident teenager who was well liked by his classmates and had recently had his first kiss and fallen head over heels in love. On August 22, 2015, he was thirteen years old when he was hit by a car at a junction while cycling to an athletics training session. Tommy-Boy was cycling very fast, but was not looking at the road. Instead, he was busy on his cell phone putting together a playlist on Spotify because he was in charge of the music for his sister Summer's birthday party later that evening. The driver of the car was traveling with four children when she hit him, through no fault of her own. She hadn't been driving too fast, but couldn't react on time when the cyclist appeared out of nowhere in front of her.

Today, Tommy-Boy's father, Michael Kulkens, travels around to schools telling the story of his son's accident. He has also started a campaign to ban the use of smartphones in traffic. He begins each talk that he gives with the track "Butterflies" by DJ Tiësto, the song Tommy-Boy was listening to at the time of the accident. The traffic situation at the junction was altered afterwards, with wooden poles installed to ensure that cyclists have to slow down before crossing the road.

When we think of concentration, we usually associate it with working or studying, but it is also crucial in the outside world, especially in traffic. The number of traffic accidents and fatalities has risen sharply in the Netherlands over the past few years and continues to rise. In 2017, recovery services had to be called out for 26,000 accidents on major roads, a rise of over 27 percent compared with the figures for 2013. On one particular stretch of highway on the A12 near Arnhem, the increase was 91 percent. And if you think that's bad, on the A73 near Nijmegen the figure rose by a staggering 131 percent. This rise in the number of accidents is unusual because the trend up to 2013 (20,000 accidents) was a downward one, and the quality of roads and cars has improved over the past few years.

One explanation for the rise in the number of accidents is a lack of concentration due to the use of smartphones in traffic. Various studies have been carried out into the cell phone use of individuals involved in traffic accidents. One of these studies investigated a total of 699 accidents and discovered that in 24 percent of these cases there was evidence of the use of a cell phone in the ten-minute period prior to the accident. Of course, there is no direct causal connection here, as there is no evidence that using a cell phone caused any of these accidents. It could simply be the case that people who drive dangerously are more inclined to use their cell phones while driving.

To investigate a possible causal connection, an experimental situation needs to be set up in which the driving behavior of test subjects without a smartphone can be compared with that of test subjects using a smartphone. This kind of experiment is far too dangerous to be carried out on the public road, but these days there is an excellent alternative: the driving simulator. Tools like these are becoming increasingly better at simulating driving behavior and conditions in the real world, and the results of the studies in which they are used are often very revealing.

Researchers at the University of Utah decided to compare the driving behavior of two groups of test subjects. One group had to conduct a conversation about a topic of their own choice over the phone with a research assistant, and the other group was asked to drink orange juice mixed with vodka so that they would have too much alcohol in their blood. Both groups then had to drive on a virtual multilane highway and stay behind the car driving in front of them, which used its brakes at random moments. When the sober test subjects were not talking on the phone, they stepped on the brakes a lot quicker in reaction to the car in front of them than when they were using their phone. The inferior driving performance while conducting a telephone conversation turned out to be on a par with the poor driving behavior of the intoxicated test subjects. It is worth reminding ourselves that the fine for holding a cell phone while driving is a lot lower than when you are caught behind the wheel with too much alcohol in your blood. Furthermore, the test drivers exhibited more erratic driving behavior when speaking on their cell phone. For example, they were much slower to accelerate again after having to apply the brakes than when they had not been talking on the phone. This kind of driving behavior is one of the main causes of congestion, especially in heavy traffic.

It is high time that we did something about improving the concentration of drivers on our roads. This will require input from relevant authorities and road users alike. In the Netherlands, it is not permitted to hold a cell phone in your hand and make a call or send a text message while driving (i.e., while the wheels are in motion). The important detail here is when you are allowed to hold the device and when you are not. Hands-free use of your phone is allowed at all times. And when your car comes to a halt, whether in a traffic jam or at a red light, you may take your cell phone in your hand if you wish. So although it is not permitted to hold your cell phone and make a phone call while driving, you may use it while the car is moving to send a text message if it is sitting in the car kit. This is a bit weird. If you make a phone call while driving, you will at least have your eyes on the road, but if you use your smartphone to look up a telephone number while driving, you have to avert your eyes from the road, if only for a moment or two. However, you and your car could cover quite a distance in the meantime. At a speed of 30 mph, you will travel twenty-six yards in three seconds, and at 75 mph this increases to 110 yards.

Hands-free phone calls are no safer than when you are holding your cell phone in your hand, but current legislation takes a different view. This is a mistake. A driver's attention will be distracted just as much when making a hands-free call as when they are holding the phone in their hand. The negative consequences of conducting a phone call while driving have nothing to do with physically holding the device, but are related instead to the driver's level of attention. When you are making a call, you have to use your working memory, meaning that you have to use the phonological loop to formulate a sentence, and in the meantime your attention is required to drive the car. If too much attention is required to retain information in the working memory, as is the case with a telephone call, there will not be enough attention left over to be able to keep a good eye on the road in front of you (even when your eyes are literally on the road).

The Eye Movements of a Formula 1 Driver

A TV program featuring the Formula 1 driver Nico Hülkenberg recently demonstrated just how important our eyes are when we are out on the road. Given the speed at which they zip around a track, racing drivers have to be able to react incredibly quickly to the information in their immediate environment. To find out how they experience the world of speed, Hülkenberg agreed to appear on a TV program in which his eye movements would be monitored using a new kind of technology while he sped around the racetrack.

Now, in my work as a scientist I have seen my fair share of eye movements, but I could barely believe the results of Nico Hülkenberg's test when I saw them. He made extremely efficient eye movements and not one of them appeared to be surplus to requirements. Countless hours of training behind the wheel have obviously taught him how to look exactly and only where he needs to look. In fact, you could say that Hülkenberg actually looks into the future because he concentrates much of his vision on the next apex (the highest point in a turn). He also processes all of the visual information he receives very quickly. The readings show that, on average, it takes him less than 100 milliseconds to put his foot to the floor the moment the lights go green at the start of a race.

The effects of years of training can also be seen in the way Hülkenberg looks in his wing mirror when exiting the pit lane. He glances at the mirror for only 100 milliseconds, which is the absolute minimum amount of time required to gather information during a fixation (the length of time that the eye is still). You or I would have to look a lot longer to be able to identify an object during a fixation.

Hülkenberg demonstrates something that every driving instructor tries to make clear in lesson number one: how important it is to know where to look when you are driving. We should never underestimate how much of a challenge it is to maneuver safely in traffic. Driving a car requires us to use our attention very efficiently, so we should appreciate just how precious a commodity our attention is.

Automatism in Traffic

When my father decided to teach me the basic principles of driving a car, he actually found it very difficult to do so. He had spent so much time out on the road that driving had become almost an automatic activity for him—an automatism, in other words. We can develop an automatism because our brain also has an implicit memory, one whose contents we are not consciously aware of. It contains, for example, motor actions that you have performed so often you do not need to think about them anymore. When you are learning how to drive, you have to think about each and every action, pressing the clutch down with your foot, pressing the brake, changing gears, and then putting your foot on the accelerator. At first, you are very aware of all these motor actions, but the more you perform them, the less you have to think about them, because they are stored in your implicit memory. Subsequently, you do not need to pay any attention to how you change gears because the action no longer requires the use of the working memory.

This means that driving is primarily an automatic activity. However, there is still a big difference between driving in a sleepy village on a Sunday afternoon and tackling a major intersection on a highway during rush hour on a Monday morning when you have to keep an eye on the traffic around you while trying to take the correct exit. You will have little trouble conducting a conversation with your fellow passengers while driving around your sleepy village, but trying to do so on a busy highway is a different matter altogether. In that case, you need to use all of your available attention to arrive safely at your destination. I can still remember the uneasy silence in our car whenever my father was driving on the ring road around Paris on our way to the south of France. Pops was not to be distracted from his task (and still we always managed to get lost).

On the one hand, it is a good thing that driving can become an automatism, because it means you can pay more attention to the road and to the traffic around you. On the other hand, it can also be dangerous to drive on autopilot because you are more likely to give in to the temptation to use your phone. Being able to react adequately to the car in front of you jamming on the brakes requires your full attention, regardless of whether you are a novice driver or have spent years driving a truck for a living. But does that mean all forms of communication should be forbidden while you are driving?

What about conducting a conversation with your fellow passengers? That requires the use of the working memory too, doesn't it? Yes, it does, but there are many different kinds of conversations you can conduct while driving and not all of them pose a threat to your safety. When test subjects in a driving simulator were asked to conduct a conversation with a friend about times in their lives when they found themselves in great danger, the number of driving errors was highest when the conversation was conducted over the phone. When their friend was sitting next to them in the passenger seat, the conversation sometimes switched to the traffic on the road, and the way in which it was conducted was adjusted when the traffic situation demanded more attention from the driver. In that case, both the driver and their conversation partner spoke less and used more straightforward language. The conversation partner appeared to take the driver's situation into consideration because they too were aware of the traffic situation and could see when the driver needed to pay more attention to the road.

This mechanism does not apply when the conversation takes place over the phone. After all, the driver's conversation partner does not have any visual information about the traffic situation and will continue talking in the same manner when the driver reaches Paris, for example, and needs to focus every single ounce of his or her attention on the city's notorious ring road. In fact, the conversation can even become a little awkward if one of the two—in this case the driver—starts speaking slower and sounding somewhat distracted. When the conversation remains complex, regardless of the traffic situation, the driver will have less attention available for watching the road, thereby increasing the risk of an accident.

This risk is lower when you are speaking to someone sitting next to you in the car than when you have them on the telephone. And when your conversation partner is actually physically present, you can see how they react to what you are saying. We garner a lot of information from the facial expressions of our conversation partners. Think of the trouble we often have trying to decipher the emojis we receive in text messages. Conversations are easier when we can use the sound of someone's voice to decipher what they are saying, but the process is even easier when we can see their face. Okay, you do have to take your eyes off the road for a moment to look at your passenger, but the disadvantages associated with having to do this are nothing compared with the consequences of the alternative. When we cannot see our conversation partner's face, we imagine it in our heads instead and this requires the use of our visual working memory—precisely the part of the working memory that is crucial when we are driving a car.

An analysis of traffic accidents in Spain between 1990 and 1999 showed that the presence of a passenger in a car actually reduced the chances of the driver having an accident. There are many possible explanations for this, of course, but one of them suggests that driving should become a joint activity. A passenger can alert the driver to potentially dangerous situations and share his or her experiences with them too (also known as *shared awareness*). Of course, this does not apply to every single conversation in a car. When the passengers are children or the conversation partner is of the "backseat driver" kind (like my mother on the ring road in Paris), it is unlikely to increase your collective safety. If nothing else, this theory is an interesting deviation from the idea that you should avoid all forms of distraction when you are driving.

Pedestrians and Their Poor Concentration

In February 2017, the Dutch town of Bodegraven saw the premiere of a strip of red and green LED lights sunk into the pavement next to a set of traffic lights, installed especially for pedestrians who, because they cannot tear their eyes away from their smartphones, end up walking out onto the road even though the lights are red. The aim of this innovation is to warn smartphone-toting pedestrians that they are approaching a traffic light so that they can cross the road safely. The plan is to install the LED lighting in pavements first, and then to explore the option of using it on bicycle paths too. Of course, the initiative is not without its critics. What about pedestrians who are colorblind? Will the strips have to be installed on every pavement just because pedestrians expect them to be there? Isn't it the pedestrian's own responsibility to look out for themselves? By doing this, aren't we just condoning the use of smartphones while walking?

Despite all of these criticisms, there was a very good reason for developing the light strip. Take the situation in Japan, for example, where so many people end up falling off train and metro platforms because they are staring at their phones that there is even a term in the Japanese language to describe it: *aruki sumaho,* or "walking with a smartphone." In 2013 alone, thirty-six people were hospitalized in Tokyo as a result of *aruki sumaho.* The situation becomes even worse when smartphone-wielding commuters are wearing headphones. In that case, their awareness of their surroundings is reduced to almost zero.

The internet is full of clips of people stumbling into fountains, walking into lampposts, and falling into ditches because their eyes (and ears) are glued to their phones. One of the most notorious clips features a woman by the name of Cathy Cruz Marrero, also known as, "The Fountain Girl." In 2011, she was walking through a shopping mall while busy using her cell phone when she fell straight into a fountain. The incident was captured on camera, posted online, and instantly went viral. She became an internet celebrity overnight and even appeared on a number of TV shows. However, she then threatened to sue the shopping mall because no one had come to her assistance when she was lying soaked in the fountain and also because the security personnel had posted the video online. Ultimately, the shopping mall didn't have to appear in court, although the employee who posted the video on the internet did end up getting fired.

In some cities in the United States, pedestrians have to pay a fine if they end up straying off the sidewalk and onto the road because they are too busy texting on their cell phone. In Fort Lee in New Jersey, for example, the fine is a hefty $54. In the state of Utah, there is a $50 fine for not paying attention when walking across a railroad crossing. In New York, pedestrians are involved in 52 percent of all traffic accidents in the city, and the ensuing costs are estimated to be as high as $1 billion each year. An observation study recently carried out in New York focused on the city's ten worst accident spots between 1995 and 2009. It revealed that out of the 3,500 pedestrians observed, one in four was busy on their cell phone while crossing the road.

Distracted pedestrians tend to walk slower and to zigzag across the sidewalk. This applies especially to children, who are among the most vulnerable when it comes to negotiating traffic. In a study using a simulator, seventy children were asked to cross a street twelve times—six times while busy on their cell phone and six times without any distractions. The results were quite shocking. When the children were distracted, they were involved in more accidents and near misses with cars. They also took more risks when crossing the street because they misjudged the proximity of the oncoming traffic and often stepped out onto the street far too late. It did not make any difference how experienced the children were at using their cell phone or crossing the street on their own. Crossing a road is a complicated cognitive process, particularly for children and the elderly. To cross safely, you have to be able to judge not only the proximity of an oncoming car but also its speed and the chances of it slowing down or accelerating. This is why people who are unable to perceive movement (as a result of brain damage or poor vision, for example) have great difficulty crossing a road safely. There is a good reason why helping an old lady to cross the road is often cited as the ultimate good deed.

The wearing of headphones also represents a major risk because you need to use all of your senses when negotiating traffic. For example, you use your hearing to judge the speed and acceleration of cars, but when you are wearing headphones this information is excluded, thereby negating half of the combination of auditory and visual information. Our brain integrates sounds and images so that it knows which sound belongs to which image, a process known as *multisensory integration*.

Multisensory integration is important because we react a lot quicker to integrated signals than to signals that are received simultaneously but are not integrated. In this case, the idea that "the whole is greater than the sum of its parts" really does apply. The entire multisensory integration system is rendered redundant when you use headphones while walking or cycling. A report on traffic accidents in the United States revealed that between 2004 and 2011 there were 116 accidents (70 percent of which were fatal) in which the pedestrian involved was wearing headphones. These accidents could have been prevented if the multisensory integration system had been allowed to work as it was designed to.

Neuromarketing Nonsense

Many different products have been launched with the aim of curtailing the use of smartphones in traffic. In 2017 in the Netherlands, for example, the telephone company KPN, working in unison with the Dutch safety authorities and the lock experts Axa, introduced a bicycle lock that prevents cyclists from using apps on their smartphone while cycling. The Safe Lock, as it is known, is connected to an app that blocks the user's access to the internet when they unlock their bicycle. The only service they can access is the emergency number. The lock retails at around 100 euro.

The examples already discussed in this chapter show that a product like the one above is clearly a very good idea. It is such a pity then that the launch of this lock was accompanied by an overly extravagant press release about a study purporting to prove the usefulness of the product. KPN had asked a neuromarketing firm to source an assortment of brain images to support the benefits of the product. Now, whenever I hear the term "neuromarketing," it always makes the scientific hairs stand up on the back of my neck.

KPN probably paid top dollar for this study, the quality of which is so poor, however, that if a student of mine produced a paper of similar quality I would have no hesitation in showing them the door. The study was of youths aged between twelve and eighteen who were monitored using electroencephalography (EEG) while performing a simple task on a computer. Electroencephalography is a method for recording electrical signals in the brain through electrodes placed on the forehead. During the task, the test subjects received intermittent notifications of incoming messages from Snapchat, Instagram, or WhatsApp. The study concluded that a person's safety can be jeopardized when they receive messages on their cell phone while cycling a bike. The press release for the new lock stated that, "research has shown that not only does the use of a smartphone while cycling cause the cyclist to become distracted, but even the sounds or vibrations associated with notifications can lead to a reduced level of attention in traffic."

There are a number of glaring problems with this study. First of all, you don't need to use images from a brain scan to measure distraction. All you have to do is study the person's behavior. The researchers defended their choice of method by stating that, "the results need to appeal to the imagination in order to get the message across in a certain way." So, according to this neuromarketing firm, because people nowadays are becoming more and more interested in neuroscience, it would be more useful to show them images of brain scans than to formulate specific research questions. Okay, the scans did show that the brain reacts to message notifications. But is there anything new about that? Of course not. After all, when you hear a sound your brain processes that too, so it is no surprise to see the resulting activity on a brain scan. What's important is to establish what influence that activity has on the test subject's behavior. And the only way to do that is to study their behavior.

Secondly, the differences in behavior recorded were not statistically reliable. So there was no real evidence that the notifications actually distracted the test subjects. This comes as no surprise either when you look at the task that the neuromarketing firm used. It was so simple that the test subjects were even inclined to nod off. In their case, all the notifications did was wake them up again and prompt them to react. Nonetheless, a lock like the one described above is certainly not a bad idea. But if we look at the accidents that we have discussed already in this chapter, we will see that the danger lies not in the notifications themselves but in the way in which we use our cell phones. When you are out walking or cycling your bike, you hear sounds all around you—people calling out to each other, someone shutting their front door, a bleep from your own telephone. In the case of such distractions, we have no problem refocusing our attention on our surroundings or the traffic in which we happen to be moving. The danger in receiving notifications on your cell phone lies in the way you react when you know you have a new message. This would not present a problem if we were always able to resist the temptation to immediately pull out our phone. But the fact is that it is often a very real problem, because many of us find it difficult to resist that temptation. You simply can't wait to find out what the message says ("mental itch"). Your hunger for new information and your eagerness to know what's in the message causes you to react to the notification by reaching for your cell phone, reading the message, and maybe even sending an immediate reply. And that's where the real danger lies, because your attention will no longer be on the road you are traveling on. All in all, I think you'll agree that you don't need an expensive study to arrive at that conclusion.

Solutions

There have been many public campaigns aimed at making road users more aware of the dangers of using a smartphone in traffic, but none appear to have been very effective. A recent government campaign in the Netherlands called "bike mode" (for staying offline in traffic) met with very little success in any case. Various apps have tried offering time locks so that users can decide for themselves how much time they want to spend on or offline. A Dutch company has even invented a hexagonal metal box in which you can place your cell phone when you don't want to be disturbed while driving. The box is lined with a material that blocks electromagnetic radiation and can hold up to six smartphones, so it can also be used by people who don't want to be interrupted when they are holding a meeting. Today, both Android and Apple have telephones with a mode that disables certain functions when the device is located inside a moving vehicle.

However, innovations like these are still finding it hard to catch on, so it is probably time for governments everywhere to start regulating the use of smartphones in traffic. In 2016, the Dutch government came up with a proposal to introduce a kind of hands-free system for bicycles that would allow cyclists to use their smartphone only when they were not holding it physically in their hand (they would have to use their headphones). Other ideas included a complete ban on smartphones in cars and a mandatory driving mode.

Nothing ever came of these plans, primarily because the authorities realized that it would be almost impossible to enforce these measures on the ground, especially in the case of cars. Nowadays, many motorists use their smartphones to navigate, and it is more or less impossible to tell whether a driver is typing in the address of their desired location or sending a text message. And there is a difference too when it comes to concentration—a big difference. Though in both cases the driver's eyes are momentarily not on the road, the situation is much more dangerous when they are sending a personal message than when they are checking their route. Navigation systems do not throw up any unexpected messages or notifications. The predictability of what will appear on the screen allows you to decide for yourself when to glance at the screen and how long you will focus your attention on it.

This is very different when it comes to using social media on your smartphone. Apps like Facebook and WhatsApp are designed to grab your attention and then retain it for as long as possible. In fact, that is more or less the definition of their business model. Take the concept of a newsfeed, for example, a never-ending stream of messages tailored to match your interests. The more you use these apps, the better the underlying algorithms get to know you, and the better the program knows how to hold your attention.

In the case of text messaging, there is the additional social pressure to react as quickly as possible. When you are using your navigation system, you don't have to worry about saying something that might be interpreted wrongly, you won't end up in an argument, and you won't find yourself waiting in tense anticipation for the next message. The complete opposite, in other words, of the conversations we conduct through text message. With texting, not only do you need to devote a lot of internal attention to interpreting the written messages, but waiting in anticipation of the next message also causes you to take your eyes off the road more often and for longer. If we want to discuss the use of smartphones in cars in a meaningful manner, we need to get the facts straight first. When driving, it makes a huge difference whether you are distracted simply because you want to wipe the crumbs from your sandwich off your lap or because you are busy catching up with messages on your smartphone. Maybe all we can do is wait until the technology comes along that will allow us to catch motorists using social media when they are driving.

It is too easy, of course, to shift the responsibility for solving these problems onto our governments. Road users also have a role to play in ensuring everyone's safety. That role is difficult to define, however, because of our tendency to overestimate our own driving skills. In the aforementioned study in which intoxicated drivers were compared with their phone-wielding compatriots, the latter stated after the test that they did not find it any more difficult to drive when they were using their phone than when they were not. The results tell a different story: they clearly had more problems driving while using the phone compared to when their attention was fully focused on the road. Most of us like to believe that using our smartphone while driving is not really a problem, but that is far from the truth. We all need to realize that concentrating on the road is absolutely crucial to road safety.

All of the factors discussed in the previous chapters that have been proven to be good for your concentration on the work floor or in the lecture hall are also good for your concentration while driving: fitness, enough sleep, and taking breaks at the right moments. We need concentration for many of the things we do in our lives, and although there may be plenty of professions or situations in which concentration is not a crucial aspect, we all use our public roads in some way or other. So, good concentration is critical for each and every one of us.

6
The Future: Are We All Going Stir-Crazy?

Anyone remember the "fidget spinner" craze of 2017? There was hardly a schoolyard anywhere that wasn't littered with them. At the peak of the hype, it was almost impossible to find a toy shop where they weren't always sold out. It was the kind of craze that most people are unable to remember only a few years down the line, and in 2018 the spinners disappeared almost as quickly as they had arrived. So, for those whose memories need refreshing: a fidget spinner is a small toy with two or three prongs arranged around a central bearing. By holding it between your thumb and index finger you can flick the prongs to make them spin. For a whole school year, children everywhere busied themselves with inventing new tricks or just staring at the spinner as it spun around and around and around . . .

The story of the origins of the fidget spinner is weird and wonderful in itself. In the media, Catherine Hettinger, an American engineer, was identified as the inventor of the gadget. It was said that she came up with the idea for the fidget spinner in 1993 as a concentration aid for her hyperactive daughter. However, she had allowed the patent she had filed for her "invention" to lapse in 2005 due to a lack of funds. After the spinners had become all the rage, she was plagued by the media with questions about how she felt now that her invention had become a huge commercial success. In an interview with *Bloomberg News*, however, Catherine revealed that the spinners as everyone knew them bore little resemblance to her own plastic toy and actually used a completely different mechanism. After that, the paper trail went dead and to this day no one knows who invented the toy that enthralled millions of children for a few short months in 2017. Furthermore, given that no patent was ever filed for the toy, it is also impossible to identify the manufacturer.

The craze eventually got so out of hand that some schools even started to ban spinners in the classroom, and by May 2017, 32 percent of the 200 largest public and private high schools in the United States had done exactly that. The problem was that the spinners had become such an enormous distraction that schoolchildren were spending most of their time in class playing with the toy. The manufacturers claimed, however, that they were of great benefit to some children, and the packaging for the toy often carried the message that the spinners could help children with ADHD to concentrate. Some even claimed that they had a positive effect on children with autism and anxiety disorders. There were also plenty of parents who said that the spinners were a great help to their children, but the reaction from the scientific community was to immediately reject these claims, saying that there was no evidence for the positive effects of fidget spinners. All of the parents' upbeat comments and experiences with the toy were simply dismissed as nonsense.

It is important, of course, that manufacturers are careful about what they print on their packaging and that they don't make unfounded claims regarding the supposed benefits of their products for people with ADHD, particularly when there is no scientific evidence to back up those claims. However, it is going a little bit too far to just dismiss out of hand all of the positive stories related to a certain product. Science is a slow, lumbering process, and the spinner craze did not last long enough to be able to study it properly, let alone publish any conclusive results in a scientific journal. We simply don't know whether the use of spinners has a positive effect on concentration or not. What we do know is that the negative coverage of the toy in the media was met with heavy criticism. When the American online magazine *Vice* published an article under the heading, "Let's Investigate the Nonsense Claim that Fidget Spinners Can Treat ADHD, Autism, and Anxiety," it quickly changed it to "Fidget Spinner Manufacturers Are Marketing Their Toys as a Treatment for ADHD, Autism, and Anxiety," after receiving a huge number of complaints. *Vice* issued an apology for using the original headline partly because of the angry reactions of parents who claimed that their children benefited enormously from using the toy.

So why were the spinners so attractive to children who had difficulty concentrating? We don't all have the same powers of concentration, of course, and this is partly due to differences in the level of activity in the motor system.

In other words, some people find it more difficult to sit still than others. It is easy to spot these differences in a classroom, where you will see some children sitting perfectly still at their desks while others seem to be in a constant state of motion. People with ADHD are the worst affected, as they have great difficulty suppressing their motor system and are therefore unable to sit still or concentrate.

Fiddling with a fidget spinner fulfills the need to stay moving, as is demanded by the motor system of overactive children. And it does not interfere with their ability to listen to the teacher. In fact, requiring them to sit still would take even more effort because they would have to put all their energy into suppressing their motor system. That is why it is never a good idea to command an overactive child to sit still in class. Various studies have shown that children with ADHD perform much better when they are allowed to move while carrying out a task that requires the use of the working memory. Just being able to move appears to keep them alert. It is not difficult to imagine, therefore, that an active child will be able to concentrate much better when they are allowed to sit on an exercise ball. This only works for busy children, by the way; quieter children actually have more difficulty concentrating when they have to move around a lot.

It is worth pointing out that many of these kinds of studies are expressly carried out on children because it is easier to diagnose disorders such as ADHD at a younger age. Adults with ADHD, for example, have often learned to compensate for their problems. Nevertheless, it is quite possible that the results of these studies can also be applied to adults too.

High Sensitivity

In September 2017, a book written by Fleur van Groningen spent two weeks in the top ten of the best-selling books list in Belgium. In her book, Van Groningen describes her experiences as a highly sensitive person (HSP). Many readers found they could identify with her story. For example, in the book she explains how she tries to minimize the amount of external stimuli she is exposed to every day and how she has learned to deal with her own often overwhelming emotions. She draws a connection between HSP and what she calls the "flood of burn-outs and depression" in Western society, where the enormous amount of stimuli results in chronic overstimulation in highly sensitive people. Many people who identify themselves as HSP often complain about having problems concentrating.

The term "high sensitivity" was first introduced in 1996 by Elaine Aron. She developed a questionnaire that people could fill in to determine whether or not they were highly sensitive. Her questionnaire is not the only one floating around, and today the internet is full of self-tests you can do to see if you are highly sensitive. The problem with these tests, however, is that the questions are so vague that it is no surprise that the results indicate that 15–20 percent of the population is highly sensitive. Many of the tests are specially designed for children and include questions like whether the child has trouble sleeping, has problems going to the lavatory, or is overly active. Based on these criteria, we could almost slap a HSP sticker on every child on the planet.

According to the definition of the term, people with HSP are more easily affected by emotions, pain, pleasure, and other physical and mental sensations. It is also claimed that they react more intensely to different sensory stimuli, including sound, touch, and color. This oversensitivity makes them extra cautious in stimulant-rich environments, which in turn causes them to come across as shy. They try to avoid the hustle and bustle of urban life and even withdraw from it as much as they can. As you have probably already noticed, I try to back up the arguments in this book with thorough scientific experiments. Unfortunately, very few reliable studies have been carried out on HSP, and those that do exist rarely rise above the level of your average self-help book.

HSP is often associated with the highly competitive society in which we live, one that is constantly demanding more and more from us. Many people have difficulty coping with the overwhelming amount of stimuli they experience every day, and HSP offers a convenient explanation for the fatigue and concentration problems they have to deal with. In recent times, HSP has become something of a hype, and there are now lots of discussion groups for people who claim to suffer from it. Many of the books written about HSP refer to highly sensitive people as having a special gift because they are able to process stimuli at a much deeper level. Whereas in science each word is carefully considered and every single term clearly defined before being published, many of the claims regarding HSP are published as hard facts without their having any scientific basis whatsoever. Of course, this is not necessarily a problem when what is being claimed is helpful to some people, but it becomes more problematic when people start regarding something like HSP as an actual disorder. This can lead parents to look for "treatment" for their highly sensitive child, though the cause of the problem may be a psychiatric disorder for which there are actual, scientifically proven methods of treatment. The problems often regarded as typical of HSP are often the same ones that are associated with ADHD and autism. One of the most typical characteristics of autism, for example, is an oversensitivity to being touched by others, something that is often linked with HSP. There are effective treatments available for these disorders, but they are not used when the treatment is focused exclusively on HSP.

High sensitivity is not a psychiatric disorder, like ADHD or autism, but more of a personality trait, like the way someone is regarded as being an extravert or an introvert. Although some people get a great kick out of the hustle and bustle of city life, others prefer a quieter lifestyle. Some children cover their ears to block out the noise of a busy playground, though other kids seem to thrive on it. However, the fact that it is not officially recognized as a psychiatric disorder does not mean that there is no such thing as HSP. Given the steady increase in the number of stimuli we have to deal with on a daily basis, it is not surprising that more and more people are experiencing great difficulty with processing everything that is thrown at them. The important thing here is not so much the amount of stimuli but rather the way in which we process them. We are exposed to just as many stimuli every day via our senses, from which our attention mechanism makes a selection for further processing. There are differences in how efficiently the attention systems of different people work. The few studies into HSP that have stood up to scientific scrutiny have shown that, in the case of people who tick most of the boxes on the HSP questionnaire, certain areas of the brain react more intensely to sensory stimuli. Interestingly, these areas are also those that are responsible for attention. Although this does not prove whether or not the increased activity in the brain also leads to problems, the results do show that there are differences in how individual brains process incoming stimuli. If large doses of incoming information are a source of trouble for you, then it is certainly a good idea to try and figure out how to deal with the problem. Our brains are simply not able to process all of the stimuli we are exposed to, and our ability to concentrate is determined in part by our ability to ignore irrelevant information. No brain is able to concentrate when there is too much distraction going on around it. But we don't need to label this as some kind of disorder or condition. The simple fact of the matter is that one brain has more difficulty concentrating than the other.

The Evolution of Concentration

The story of the fidget spinners shows us that there are differences in how well people are able to concentrate and that those of us who suffer from concentration problems sometimes need to use tricks to help us concentrate, particularly in today's world with all its distractions. I recently came across an interesting article in a Dutch newspaper by Michael Pietrus, a psychologist at the University of Chicago, in which he said that, thanks to the arrival of the smartphone and social media, more and more people are behaving like they have ADHD. He doesn't say that we all actually have ADHD, but rather that our behavior is becoming very similar to that of people who do have ADHD. This may sound a bit over the top to some, but it appears that our current love for multitasking is seriously affecting our ability to concentrate on one task at a time.

But is the rise in ADHD actually related to the rise in the number of stimuli in today's world? Over the past few decades the number of cases of ADHD has risen dramatically, and a correlation has consequently been made with the ever-increasing number of stimuli. This correlation is often supported by studies of a medium that has been around a lot longer than the smartphone, one that is also a plentiful source of stimuli: the television. The library of scientific literature on the influence of TV on our development holds far more reading material than the one housing studies into the social media (admittedly a more recent phenomenon) that our children are now being exposed to from a very young age.

The television has had an interesting career up to now. When it first arrived on the scene, there was a very limited number of channels and there was no such thing as a remote control for switching quickly from one station to the next, as I have already explained. The early years were characterized by a curious phenomenon: "peak attention," a term first used by the author Tim Wu. According to Wu, the ultimate peak came on September 9, 1956, when Elvis

Presley appeared on *The Ed Sullivan Show*. In the United States, almost 83 percent of the population tuned in to witness the moment (in other words, eight out of every ten people in the United States were all doing the same thing at the exact same time: watching Elvis).

In the 1960s, some shows on American television were able to draw over 60 million viewers, the kind of figure that present-day advertisers dream about at night. The dream was short-lived, however, thanks to the arrival of the remote control and the emergence of commercial broadcasters. If the viewer became bored, even for a fleeting moment, they could immediately zap to the next channel. This development resulted in TV shows becoming increasingly fast-paced so that the viewer wouldn't even have time to think about switching channels. Everything had to happen at breakneck speed, and so the editing became more rapid and the camera positions more plentiful. Today, whenever I watch a kids' TV show, I can hardly believe the overwhelming wall of sound and images that is blasted at my children. On the other hand, whenever I show them what I used to watch on TV when I was a kid, they quickly switch off. Far too slow, they say. And then even I have to wonder how anyone ever watched programs that were so sluggish.

Many parents fear that the avalanche of images that their children are exposed to on TV these days can have a negative effect on the development of their brains. Their fears appear to be justified by studies of how long children spend watching TV and the effect it has on them. One such study was carried out by scientists at the University of Washington in Seattle. They interviewed the parents of 1,278 children aged one and 1,345 kids aged three about their children's TV viewing behavior. The children in the most extreme group watched an average of two hours of TV each day. These children were then tested again for hyperactivity when they were seven years old. The results showed that were was a strong link between how much TV they watched when they were very young and the extent of the attention problems they suffered from a few years later.

Similar studies have also revealed a link with reading problems. You have probably already guessed the problem with these studies: they represent nothing but correlations. There is no reason to assume that the children's hyperactivity was the result of watching television. It may simply be the case that children who are inclined to be hyperactive watch more television because they find it difficult to concentrate on some other activity. This is clearly articulated in the studies, but their conclusions are often taken completely out of context by the media just so that they can reveal a causal connection between hyperactivity and watching TV. An interesting hypothesis maybe, but nothing more than that.

Fortunately, evolution is a slow-moving process, and so we are unlikely to be confronted all of a sudden with a generation of children who are completely unable to concentrate for a significant length of time because of structural changes in their brains. The potential of our brains remains the same, but our surroundings are changing all the time, thanks to the increase in stimuli. Ultimately, concentration is like a muscle in that it requires training to stay strong. The more you train, the better your concentration. However, not everyone develops the same kind of muscle power, regardless of how hard they train. Some people simply have a greater capacity for building concentration muscles than others, and this doesn't change even if you keep increasing the number of stimuli. One thing that has definitely changed, however, is the amount of effort that is required from our concentration muscles, and the more stimuli you are exposed to and the more you are inclined to multitask, the more effort you will have to put in if you want to maintain your concentration. The trick is to keep the muscles as strong as possible while at the same time keeping the amount of effort required to a minimum.

Hope for the Future

Recently there has been a deluge of ex-employees from large technological firms in Silicon Valley anxious to share their woeful tales with the world regarding the direction in which our society is heading. These apparently conscience-stricken executives, all of whom have earned a fortune in the technology business, now look back with (feigned?) regret at the innovations they helped to turn into reality. For example, James Williams, a former employee of Google, talks about facing "the most challenging crisis of our time," adding that "we drag our smartphones and social media around with us everywhere. They weigh us down and reduce our chances of being happy and successful." These doom and gloom merchants are often the driving force behind new startups aimed at developing products that can shield us from social media, or they spend their time giving extremely well-paid talks on the disasters that are about to unfold.

In this book, I have tried to dispel a number of myths related to concentration. No, we don't have a shorter span of attention than goldfish. No, our IQ does not fall as a result of our liking for multitasking. On the very day that I was finishing this chapter, I found myself reading an interview with Alan Lightman, a professor at the Massachusetts Institute of Technology, in which he claimed that our creative powers have declined measurably since the 1990s because we spend too little time doing nothing. And, as is often the case with these kinds of hyped-up stories, the resulting damage is even compared to the harmful effects of smoking.

It is a real shame that important questions about our society are often met with this kind of fearmongering, the kind that leaves no room at all for nuance. It is too easy to sketch doom scenarios for our future without having to provide a single shred of evidence. There is simply no evidence, for example, that Google is making us all dumber by the minute or that the rise of the smartphone has blunted our creativity. Creativity is far too complex a thing to be able to measure it using a single test. Not to mention the fact that there are countless ways in which we can be creative, and that today the list is becoming longer, not shorter. Creativity can come in the form of an ingenious computer code, a modern work of art, or a new digital application. With this in mind, we can conclude that the standard method for testing creativity (the Torrance Tests) is hopelessly outdated. New and important scientific and technological discoveries are being made every single day, and there is no sign that the brain is losing its ability to concentrate and be creative as a result. However, we do need to be aware that technological advances can mean that we fail to realize our full human potential and can even lead to dangerous situations, like in traffic. Fortunately, science has taught us a lot about concentration in recent years and has also provided us with the tools we need to be able to concentrate and work in an efficient and creative manner. For example, we are now more aware of the positive effect of taking breaks (preferably in natural surroundings), concentration training, meditation, and switching tasks at the right moment. And we also know how concentration can be negatively affected by multitasking and surrounding yourself with potential distractions.

Whichever tools you choose, they will not automatically provide you with better powers of concentration, because those powers depend on many other factors too. There are individual differences with regard to capacity and the length of time you can concentrate depends partly on the nature of the task at hand, and the optimal span of concentration is dependent on factors like fitness levels and fatigue. This is why self-help books that offer fixed training programs for improving your concentration do not work. Concentration is made to measure, and it is up to each individual to find out what suits them best.

In addition to the things we can do ourselves to improve our concentration, there are steps that our governments could take too. Banning LED advertising (especially the animated kind) from our public highways would improve road safety, as would a ban on the use of social media in traffic. Schools should remain ad-free zones, and firm agreements need to be made regarding the use of smartphones in the classroom. The importance of concentration is also something that could be taught in our schools. A knowledge of how concentration works and how social media try to kidnap our attention can help us to stay in charge of our own attention. Children should be taught how addictive social media can be and how distracting the notifications on their smartphones are too. The more we know about our brain and our behavior, the more we will understand how we function as human beings.

It is important that we remain aware of just how valuable concentration is to us. Of course, it is not beyond the bounds of possibility that some day we will turn our backs on social media and pay no more attention to all those meaningless likes. In the meantime, it is a reassuring and wonderful thought that we are ultimately the masters of our own attention and can decide to opt out of the attention rat race if we wish. In fact, you can defeat the mighty machine of Silicon Valley at the mere touch of a button. All you have to do is switch off your phone.

Acknowledgments

After the publication of the Dutch edition of my first book *How Attention Works* in 2016, one of the questions I was repeatedly asked was when I was going to start writing the next one. I had never really planned on writing a second book, mostly because I didn't know what I would want to write about. However, it became very clear to me from the many conversations and meetings I had with people as a result of my first book that attention was a much broader subject than I had initially suspected. Some of those conversations were with insurance brokers, bridge tenders, teachers, traffic psychologists, and people from the corporate sector. Through them I discovered that attention is a major issue in many people's lives, people who wonder what they can do to remain focused in these times of infinite distraction. Much of what you have read in this book is based on those conversations and meetings. I would like to thank everyone I met for their words of wisdom and ideas that eventually became the bones of this book.

I would never have even considered embarking upon a second book if the process of writing the first one had not been such a great source of pleasure. For that I am particularly grateful to the people at Maven Publishing, including Sander Ruys, Lydia Busstra, and Evelien Pabbruwe. That the book is legible at all is entirely down to Emma Punt, Mariska Heijmen, and Juliët Jonkers. Emma, thanks for the brainstorming sessions aimed at getting the message in the book just right.

I am very grateful to the following people for checking specific passages: Mark Nieuwenstein, Serge Dumoulin, Chris Paffen (thanks for everything), Leon Kenemans, Heleen Slagter, and Edwin Dalmaijer. They spotted mistakes that would otherwise have been overlooked, but I take all responsibility for any errors that might have crept through unnoticed.

I would like to thank everyone at the AttentionLab at Utrecht University for the terrific working atmosphere and exciting scientific discoveries we have shared over the past few years. Some would say that it is quite a challenge to write and publish a book, but for me the beating heart of my work and the real thrills still lie in the experiments we conduct. I am very grateful to the Faculty of Social and Behavioral Sciences for their trust and support. The Experimental Psychology Department at the University of Utrecht is a great place to work; the Young Academy, a club I am proud to be a member of; and Tanja Nijboer, an inspirational and at times hilarious colleague to work with.

A warm thank you to my parents and friends for all your help and camaraderie. Jannie, thank you for all the love and for giving me the freedom to set out on these little adventures of mine. And last but not least, thanks to Jasper and Merel for keeping me focused on the things that really matter.

Notes

Prologue

Concerns Regarding Information Overload Throughout History
Wellmon, C. (2012). Why Google isn't making us stupid ... or smart. Retrieved from http://www.iasc-culture.org/THR/THR_article _2012_Spring_Wellmon.php.

The Attention Economy
Crawford, M. (2015). *The world beyond your head: How to flourish in an age of distraction.* New York: Farrar, Straus, and Giroux.

The History of Commercial Public Advertising and the Rise of the First Commercial Newspapers and Radio Stations
Wu, T. (2016). *The attention merchants: From the daily newspaper to social media, how our time and attention is harvested and sold.* New York: Alfred A. Knopf.

How Streaming Services Are Changing Music
Kraak, H. (2017, 19 November). Hoe streamingdiensten als Spotify de muziek veranderen [How streaming services like Spotify

are changing music]. *de Volkskrant.* Retrieved from https://www
.volkskrant.nl/cultuur-media/hoe-streamingdiensten-als-spotify
-de-muziek-veranderen~b10594f9/.

The Inability to Spot a Clown on a Unicycle
Hyman, I. E., Boss, S. M., Wise, B. M., McKenzie, K. E., & Caggiano, J. M. (2010). Did you see the unicycling clown? Inattentional blindness while walking and talking on a cell phone. *Applied Cognitive Psychology, 24,* 597–607.

Other Field and Experimental Studies into the Effect of Mobile Phones on Walking Behavior
Hatfield, J., & Murphy S. (2007). The effects of mobile phone use on pedestrian crossing behavior at signalized and unsignalized intersections. *Accident Analysis and Prevention, 39*(1), 197–205.

Nasar, J., Hecht, P., & Wener, R. (2008). Mobile phones, distracted attention, and pedestrian safety. *Accident Analysis and Prevention, 40*(1), 69–75.

The Causes of Accidents Involving Pedestrians
Nasar, J., & Troyer, D. (2013). Pedestrian injuries due to mobile phone use in public places. *Accident Analysis and Prevention, 57*(1), 91–95.

Watson's Manifesto
Watson, J. B. (1913). Psychology as the behaviorist views it. *Psychological Review, 20*(2), 158–177.

Pavlov and Conditioning
Pavlov, I. P. (1927). *Conditioned reflexes.* Oxford, England: Oxford University Press.

Free Will and Behaviorism
Ferster, C. B., & Skinner, B. F. (1957). *Schedules of reinforcement.* Upper Saddle River, NJ: Prentice-Hall.

Addiction and Rats
Wise, R. A. (2002). Brain reward circuitry: Insights from unsensed incentives. *Neuron, 36*(2), 229–340.

Man as Conditioned Dog and the Mobile Phone
Stafford, T. (2006, September 19). Why email is addictive (and what to do about it). *Mind Hacks.* Retrieved from https://mindhacks.com/2006/09/19/why-email-is-addictive-and-what-to-do-about-it/

The Mental Itch to Check Your Email
Levitin, D. J. (2014). *The organized mind: Thinking straight in the age of information overload.* New York, NY: Plume/Penguin Books.

The Importance of Attention in Health Care
Klaver, K., & Baart, A. (2011). Attentive care in a hospital: Towards an empirical ethics of care. *Medische Antropologie, 23*(2), 309–324.

Johansson, P., Oléni, M., & Fridlund, B. (2002). Patient satisfaction with nursing care in the context of health care: A literature study. *Scandinavian Journal of Caring Sciences, 16*(4), 337–344.

Radwin, L. (2000). Oncology patients' perceptions of quality nursing care. *Research in Nursing & Health, 23*(3), 179–190.

Suggestion That Health Care Workers Should Focus Only on Essential Forms of Care
van Jaarsveld, M. (2011, 21 June). Zorg is overheidstaak, aandacht geven niet [Health care is a government task, not giving attention]. *Trouw.* https://www.trouw.nl/opinie/zorg-is-overheidstaak-aandacht-geven-niet~ba6780ed.

The Increase in Available Information

Alleyne, R. (2011, 11 February). Welcome to the information overload—174 newspapers a day. *Telegraph.* https://www.telegraph.co.uk/news/science/science-news/8316534/Welcome-to-the-information-age-174-newspapers-a-day.html.

Chapter 1: Why Is It Difficult to Concentrate?

The Philosophy behind Our External Memory

Clark, A., & Chalmers, D. J. (1998). The extended mind. *Analysis, 58*(1), 7–19.

Embodied Cognition

Rowlands, M. (2010). The mind embedded. In: *The new science of the mind: From extended mind to embodied phenomenology* (pp. 1–23). Cambridge, MA: MIT Press.

Shapiro, L. (2011). *Embodied Cognition.* New York, NY: Routledge.

Improving Your Memory by Acting Out a Story

Scott, C. L., Harris, R. J., & Rothe, A. R. (2001). Embodied cognition through improvisation improves memory for a dramatic monologue. *Discourse Processes, 31*(3), 293–305.

Storing Information by Taking Notes

Mueller, P. A., & Oppenheimer, D. M. (2014). The pen is mightier than the keyboard: Advantages of longhand over laptop note taking. *Psychological Science, 25*(6), 1159–1168.

The Iconic Memory

Sperling, G. (1960). Negative afterimage without prior positive image. *Science, 131,* 1613–1614.

The Echoic Memory

Sams, M., Hari, R., Rif, J., & Knuutila, J. (1993). The human auditory sensory memory trace persists about 10 sec: Neuromagnetic evidence. *Journal of Cognitive Neuroscience, 5*, 363–370.

The Effect of Context on Long-Term Memory

Godden, D. R., & Baddeley, A. D. (1975). Context-dependent memory in two natural environments: On land and underwater. *British Journal of Psychology, 66*(3), 325–331.

Individual Differences in the Capacity of the Working Memory

Jarrold, C., & Towse, J. N. (2006). Individual differences in working memory. *Neuroscience, 139*(1), 39–50.

The Necessity of Repeating Information in the Working Memory

Peterson, L. R., & Peterson, M. J. (1959). Short-term retention of individual verbal items. *Journal of Experimental Psychology, 58*(3), 193–198.

Chunking Information (Such As Race Times)

Ericsson, K. A., Chase, W. G., & Faloon, S. (1980). Acquisition of a memory skill. *Science, 208*(4448), 1181–1182.

Postal Codes in the United Kingdom

Royal Mail Group (2016, June 4). Royal Mail reveals why we never forget a postcode, 57 years after its introduction. Retrieved from https://www.royalmailgroup.com/en/press-centre/press-releases/royal-mail/royal-mail-reveals-why-we-never-forget-a-postcode-57-years-after-its-introduction/.

Baddeley's Model of the Working Memory

Baddeley, A. D., & Hitch, G. J. (1974). Working memory. In G. H. Bower (Ed.), *The psychology of learning and motivation* (pp. 47–89). New York, NY: Academic Press.

Baddeley, A. D., & Hitch, G. J. (1994). Developments in the concept of working memory. *Neuropsychology, 8*(4), 485–493.

Baddeley, A. D. (2003). Working memory: Looking back and looking forward. *Nature Reviews Neuroscience, 4*, 829–839.

Articulatory Suppression and the Phonological Loop

Baddeley, A. D., Thomson, N., & Buchanan, M. (1975). Word length and the structure of short-term memory. *Journal of Verbal Learning and Verbal Behavior, 14*, 575–589.

Mental Rotation

Shepard, R. N., & Metzler, J. (1971). Mental rotation of three-dimensional objects. *Science, 171*(3972), 701–703.

Better Mental Rotation in Athletes and Musicians

Pietsch, S., & Jansen, P. (2012). Different mental rotation performance in students of music, sport and education. *Learning and Individual Differences, 22*(1), 159–163.

The Effect of Physical Exercise on Mental Rotation

Moreau, D., Mansy-Dannay, A., Clerc, J., & Guerrién, A. (2011). Spatial ability and motor performance: Assessing mental rotation processes in elite and novice athletes. *International Journal of Sport Psychology, 42*(6), 525–547.

The Difference between Sexes in Mental Rotation

Quinn, P. C., & Liben, L. S. (2008). A Sex Difference in Mental Rotation in Young Infants. *Psychological Science, 19*(11), 1067–1070.

Complex Items in the Visuospatial Sketchbook

Luck, S. J., & Vogel, E. K. (1997). The capacity of visual working memory for features and conjunctions. *Nature, 390*(6657), 279–281.

Brain Damage and the Wisconsin Card Sorting Task

Milner, B. (1963). Effect of different brain lesions on card sorting. *Archives of Neurology, 9*(1), 90–100.

The Effect of Age on Performance on the Wisconsin Card Sorting Task

Huizinga, M., & van der Molen, M. W. (2007). Age-group differences in set-switching and set-maintenance on the Wisconsin Card Sorting Task. *Developmental neuropsychology, 31,* 193–215.

The Switch in Our Brain

Corbetta, M., & Shulman, G. L. (2002). Control of goal-directed and stimulus-driven attention in the brain. *Nature Reviews Neuroscience, 3*(3), 201–215.

Missed Signs of a Planned School Shooting

Sandy Hook Promise (2016, December 2). Evan. YouTube video, 2:28. Retrieved from https://www.youtube.com/watch?v=A8syQeFtBKc.

Attention Problems and PTSD

Vasterling, J. J., Brailey, K., Constans, J. I., & Sutker, P. B. (1998). Attention and memory dysfunction in posttraumatic stress disorder. *Neuropsychology, 12*(1), 125–133.

Honzel, N., Justus, T., & Swick, D. (2014). Posttraumatic stress disorder is associated with limited executive resources in a working memory task. *Cognitive, Affective, & Behavioral Neuroscience, 14*(2), 792–804.

Writing Your Worries Away
Ramirez, G., & Beilock, S. L. (2011). Writing about testing worries boosts exam performance in the classroom. *Science*, *331*(6014), 211–213.

The Costs of a Complicated Mathematical Equation
Ashcraft, M. H., & Kirk, E. P. (2001). The relationships among working memory, math anxiety, and performance. *Journal of Experimental Psychology: General*, *130*(2), 224–237.

Chapter 2: When and When Not to Multitask

The Blunder at the Oscars in 2017
Pulver, A. (2017, February 27). Anatomy of an Oscars fiasco: how *La La Land* was mistakenly announced as best picture. *Guardian*. Retrieved from https://www.theguardian.com/film/2017/feb/27/anatomy-of-an-oscars-fiasco-how-la-la-land-was-mistakenly-announced-as-best-picture.

Youngs, I. (2017, February 27). The woman knows who's won the Oscars ... but won't tell. *BBC News*. Retrieved from http://www.bbc.com/news/entertainment-arts-38923750.

O'Connell, J. (2017, February 28). Was smartphone distraction the cause of the Oscars error? *Irish Times*. Retrieved from http://www.irishtimes.com/culture/was-smartphone-distraction-the-cause-of-the-oscars-error-1.2992296.

The Activity in the Brain During Multitasking
Clapp, W. C., Rubens, M. T., & Gazzaley, A. (2010). Mechanisms of working memory disruption by external interferences. *Cerebral Cortex*, *20*(4), 859–872.

Task-Switching
Rubinstein, J. S., Meyer, D. E., & Evans, J. E. (2001). Executive control of cognitive processes in task switching. *Journal of Experimental Psychology: Human Perception and Performance, 27*(4), 763–797.

The Effect of Choosing to Switch Tasks
Leroy, S. (2009). Why is it so hard to do my work? The challenge of attention residue when switching between work tasks. *Organizational behavior and human decision processes, 109*(2), 168–181.

Percentage of Youth Who Media-Multitask versus the Rest of the Population
Carrier, L. M., Cheever, N. A., Rosen, L. D., Benitez, S., & Chang, J. (2009). Multitasking across generations: Multitasking choices and difficulty ratings in three generations of Americans. *Computers in Human Behavior, 25*, 483–489.

Cognitive Flexibility and Media Users
Ophir, E., Nass, C., & Wagner, A. D. (2009). Cognitive control in media multitaskers. *Proceedings of the National Academy of Sciences, 106*(37), 15583–15587.

Estimating the Capacity to Multitask
Sanbonmatsu, D. M., Strayer, D. L., Medeiros-Ward, N., & Watson, J. M. (2013). Who multi-tasks and why? Multi-tasking ability, perceived multi-tasking ability, impulsivity, and sensation seeking. *PLOS ONE, 8*(1).

Task-Switching among Office Workers
Mark, G., Gonzales, V. M., & Harris, J. (2005). No task left behind? Examining the nature of fragmented work. In: *Proceedings of the SIGCHI Conference on Human Factors in Computing Systems* (pp. 321–330). New York, NY: ACM.

Wajcman, J., & Rose, E. (2011). Constant connectivity: Rethinking interruptions at work. *Organization Studies, 32*(7), 941–961.

Jackson, T., Dawson, R., & Wilson, D. (2002). Case study: Evaluating the effect of email interruptions within the workplace. In: *Proceedings of EASE 2002: 6th International Conference on Empirical Assessment and Evaluation in Software Engineering* (pp. 3–7). Keele, UK: Keele University.

References to Multitasking and Task-Switching on the Work Floor

Gazzaley, A., & Rosen, L. D. (2016). *The distracted mind: Ancient brains in a high-tech world.* Cambridge, MA: MIT Press.

Experiences of Workers Who Are Prone to Task-Switching

Mark, G., Gudith, D., & Klocke, U. (2008). The cost of interrupted work: More speed and stress. *Proceedings of the SIGCHI conference on Human Factors in Computing Systems* (pp. 107–110). New York, NY: ACM.

The Effects of Multitasking on Learning

Foerde, K., Knowlton, B. J., & Poldrack, R. A. (2006). Modulation of competing memory systems by distraction. *Proceedings of the National Academy of Sciences, 103*(31), 11778–11783.

Study of the Effect of Multitasking on IQ

Wilson, G. (2010, January 16). Infomania experiment for Hewlett-Packard. Retrieved from http://www.drglennwilson.com/ Infomania_experiment_for_HP.doc.

retrospectacle (2007, February 27). Hewlett Packard "infomania" study pure tripe, blogs not. *ScienceBlogs.* Retrieved from http:// scienceblogs.com/retrospectacle/2007/02/27/hewlett-packard -infomania-stud/.

Students' Powers of Concentration

Rosen, L. D., Carrier, L. M., & Cheever, N. A. (2013). Facebook and texting made me do it: Media-induced task-switching while studying. *Computers in Human Behavior, 29*(3), 948–958.

Distractions While Studying

Judd, T. (2014). Making sense of multitasking: The role of Facebook. *Computers & Education, 70*, 194–202.

Rosen, L. D., Carrier, L. M., & Cheever, N. A. (2013). Facebook and texting made me do it: Media-induced task-switching while studying. *Computers in Human Behavior, 29*(3), 948–958.

Wang, Z., & Tchernev, J. M. (2012). The "myth" of media multitasking: Reciprocal dynamics of media multitasking, personal needs, and gratifications. *Journal of Communication, 62*(3), 493–513.

Correlations between Multitasking and Exam Results

Levine, L. E., Waite, B. M., & Bowman, L. L. (2007). Electronic media use, reading, and academic distractibility in college youth. *Cyberpsychology & Behavior, 10*(4), 560–566.

Clayson, D. E., & Haley, D. A. (2013). An introduction to multitasking and texting: Prevalence and impact on grades and GPA in marketing classes. *Journal of Marketing Education, 35*(1), 26–40.

Burak, L. (2012). Multitasking in the university classroom. *International Journal of Scholarship of Teaching and Learning, 6*(2), 8.

The Correlation between Pulling Teeth and Memory

Mensen zonder tanden hebben slechter geheugen [People without teeth have worse memory] (2004, October 28). *NU.* http://www.nu.nl/algemeen/433197/mensen-zonder-tanden-hebben-slechter-geheugen.html.

Experimental Study into Media Usage During Lectures and Study Time

Wood, E., Zivcakova, L., Gentile, K., De Pasquale, D., & Nosko, A. (2011). Examining the impact of off-task multi-tasking with technology on real-time classroom learning. *Computers & Education, 58*(1), 365–374.

Kuznekoff, J. H., & Titsworth, S. (2013). The impact of mobile phone usage on student learning. *Communication Education, 62*(3), 233–252.

Bowman, L. L., Levine, L. E., Waite, B. M., & Gendron, M. (2010). Can students really multitask? An experimental study of instant messaging while reading. *Computers & Education, 54*(4), 927–931.

Multitasking in the Netherlands

Voorveld, H. A. M., & van der Goot, M. (2013). Age differences in media multitasking: A diary study. *Journal of Broadcasting and Electronic Media, 57*(3), 392–408.

Listening to Music at Work

R2 Research B. V. (2012, September 18). Randstad: werknemers productiever door muziek [Randstad: Music makes employees more productive]. Retrieved from https://www.slideshare.net/mennourbanus/randstad-werknemers-productiever-door-muziek.

Don, C. (2017, August 30). Word je productiever van muziek luisteren tijdens werk? [Does listening to music make you more productive while working?] *NRC*. Retrieved from https://www.nrc.nl/nieuws/2017/08/30/nooit-opereren-zonder-muziek-12746134-a1571629.

Ten Have, C. (2012, October 16). Op de werkvloer werkt Adele het best [Adele works best in the workplace]. *de Volkskrant*. Retrieved from https://www.volkskrant.nl/nieuws-achtergrond/op-de-werkvloer-werkt-adele-het-best~b313d381.

Haake, A. B. (2011). Individual music listening in workplace settings: An exploratory survey of offices in the UK. *Musicae Scientiae*, *15*(1), 107–129.

Supermultitaskers

Watson, J. M., & Strayer, D. L. (2010). Supertaskers: Profiles in extraordinary multitasking ability. *Psychonomic Bulletin & Review*, *17*(4), 479–485.

Chapter 3: The Sender

The Starting Procedure for Speed Skating

Dalmaijer, E. S., Nijenhuis, B., & Van der Stigchel, S. (2015). Life is unfair, and so are racing sports: Some athletes can randomly benefit from alerting effects due to inconsistent starting procedures. *Frontiers in Psychology*, *6*, 1618.

The Accident at the Ketelbrug Bridge

Brugwachter Ketelbrug vrijuit na bizar ongeval [Ketelbrug bridge-keeper free after a bizarre incident] (2008, October 22). *Het Parool*. Retrieved from https://www.parool.nl/binnenland/brugwachter -ketelbrug-vrijuit-na-bizar-ongeval~a38518/.

ANP (2011, April 29). OM vervolgt wachter Ketelbrug tóch [Public Prosecution Service continues prosecution of Ketelbrug bridge-keeper]. *De Volkskrant*. Retrieved from https://www.volkskrant .nl/nieuws-achtergrond/om-vervolgt-wachter-ketelbrug-toch ~beb07f9c/.

Rechtbank wil reconstructive ongeval Ketelbrug [Court wants Ketelbrug accident reconstruction] (2008, March 25). *De Volkskrant*. Retrieved from https://www.volkskrant.nl/binnenland/ rechtbank-wil-reconstructie-ongeval-ketelbrug~a963733/.

The Alertness of Radar Personnel

Mackworth, N. H. (1948). The breakdown of vigilance during prolonged visual search. *Quarterly Journal of Experimental Psychology, 1,* 6–21.

The Yerkes–Dodson Law

Yerkes, R. M., & Dodson, J. D. (1908). The relation of strength of stimulus to rapidity of habit-formation. *Journal of Comparative Neurology of Psychology, 18,* 459–482.

Diamond, D. M., Campbell, A. M., Park, C. R., Halonen, J., & Zoladz, P. R. (2007). The temporal dynamics model of emotional memory processing: A synthesis on the neurobiological basis of stress-induced amnesia, flashbulb and traumatic memories, and the Yerkes-Dodson law. *Neural Plasticity,* 60803.

Concerning Media Reports That We Have a Shorter Attention Span Than a Goldfish

McSpadden, K. (2015, May 4). You now have a shorter attention span than a goldfish. *Time.* Retrieved from http://time.com/3858309/attention-spans-goldfish/.

Egan, Timonthy. (2016, January 22). The eight-second attention span. *New York Times.* Retrieved from http://www.nytimes.com/2016/01/22/opinion/the-eight-second-attention-span.html.

The Original Data from the Microsoft Study

Statistic Brain Research Institute (2018, March 2). Attention span statistics. Retrieved from http://www.statisticbrain.com/attention-span-statistics/.

Why the Microsoft Report Is Nonsense

Milano, D. (2019, January 1). No, you don't have the attention span of a goldfish. *Ceros Originals.* Retrieved from https://www.ceros.com/originals/no-dont-attention-span-goldfish/.

The Memory of Goldfish

Brown, C. (2015). Fish intelligence, sentience, and ethics. *Animal Cognition, 18*(1), 1–17.

Attention Span During Lectures

Wilson, K., & Korn, J. H. (2007). Attention during lectures: Beyond ten minutes. *Teaching of Psychology, 34*(2), 85–89.

Average Video Viewing Times

Smith, A. (2015, December 2). What's the optimal length for a YouTube vs. Facebook video? *Tubular Insights.* Retrieved from https://tubularinsights.com/optimal-video-length-youtube-facebook/.

Stone, A. (2016, June 4). The lie of decreasing attention spans. LinkedIn. Retrieved from https://www.linkedin.com/pulse/lie-decreasing-attention-spans-alvin-stone.

The Potential Benefits of Waiting before Answering Messages and of Taking Breaks from Technology

Rosen, L. D., Lim, A. F., Carrier, M., & Cheever, N. A. (2011). An empirical examination of the educational impact of text message-induced task switching in the classroom: Educational implications and strategies to enhance learning. *Psicologia Ecuativa, 17*(2), 163–177.

Rosen, L. D., Carrier, L. M., & Cheever, N. A. (2013). Facebook and texting made me do it: Media-induced task-switching while studying. *Computers in Human Behavior, 29*(3), 948–958.

Rosen, L. D., Cheever, N. A., & Carrier, L. M. (2012). *iDisorder: Understanding our obsession with technology and overcoming its hold on us.* New York, NY: Palgrave Macmillan.

The Effect of Multitasking on Heart Rate

Mark, G., Wang, Y., & Niiya, M. (2014). Stress and multitasking in everyday college life: An empirical study of online activity. In:

Proceedings of the SIGCHI Conference on Human Factors in Computing Systems (pp. 41–50). New York, NY: ACM.

Videogaming and ADHD

Bioulac, S., Lallemand, S., Fabrigoule, C., Thoumy, A. L., Philip, P., & Bouvard, M. P. (2014). Video game performances are preserved in ADHD children compared with controls. *Journal of Attention Disorders, 18*(6), 542–550.

Chapter 4: The Receiver

The Daily Rituals of Geniuses

Currey, M. (2013). *Daily rituals: How artists work.* New York, NY: Alfred A. Knopf.

Resting-State Measurements

Functionele netwerken in gezonde en zieke hersenen [Functional networks in a healthy and sick brain] (2009, November 24). *Universiteit Leiden News.* Retrieved from https://www.universiteitleiden.nl/nieuws/2009/11/functionele-netwerken-in-gezonde-en-zieke-hersenen.

The Brain's Default Network

Raichle, M. E., MacLeod, A. M., Snyder, A. Z., Powers, W. J., Gusnard, D. A., & Shulman, G. L. (2001). A default mode of brain function. *Proceedings of the National Academy of Sciences, 98*(2), 676–682.

Raichle, M. E. (2015). The brain's default mode network. *Annual Review of Neuroscience, 38,* 433–447.

Daydreaming While Reading

Schooler, J. W., Reichle, E. D., & Halpern, D. V. (2004). Zoning out while reading: Evidence for dissociations between experience and metaconsciousness. In: D.T. Levitin (Ed.), *Thinking and Seeing: Visual Metacognition in Adults and Children* (pp. 203–226). Cambridge, MA: MIT Press.

Daydreaming and Happiness

Killingsworth, M. A., & Gilbert, D. T. (2010). A wandering mind is an unhappy mind. *Science, 330*(6006), 932–932.

The Relationship between Daydreaming and Cognitive Skills

Mrazek, M. D., Smallwood, J., Franklin, M. S., Baird, B., Chin, J. M., & Schooler, J. W. (2012). The role of mind-wandering in measurements of general aptitude. *Journal of Experimental Psychology General, 141*, 788–798.

Schooler, J. W., Mrazek, M. D., Franklin, M. S., Baird, B., Mooneyham, B. W., Zedelius, C., & Broadway, J. M. (2014). The middle way: Finding the balance between mindfulness and mindwandering. *The Psychology of Learning and Motivation, 60*, 1–33.

Unconscious Decisions and Their Benefits

Dijksterhuis, A., Bos, M. W., Nordgren, L. F., & van Baaren, R. B. (2006). On making the right choice: The deliberation-without-attention effect. *Science, 311*, 1005–1007.

Newell, B. R., & Shanks, D. R. (2014). Unconscious influences on decision making: A critical review. *Behavioral and Brain Science, 37*(1), 1–19.

Nieuwenstein, M., Wierenga, T., Morey, R., Wicherts, J., Blom, T., Wagenmakers, E.-J., & van Rijn, H. (2015). On making the right choice: A meta-analysis and large-scale replication attempt of the unconscious thought advantage. *Judgment and Decision Making, 10*(1), 1–17.

The Role of Nature in Refreshing Attention

Berman, M. G., Jonides, J., & Kaplan, S. (2008). The cognitive benefits of interacting with nature. *Psychological Science, 19*(12), 1207–1212.

Taylor, A. F., & Kuo, F. E. (2009). Children with attention deficits concentrate better after walk in the park. *Journal of Attention Disorders, 12*(5), 402–409.

Kaplan, R. (2001). The nature of the view from home: Psychological benefits. *Environment & Behavior, 33*(4), 507–542.

Berman, M. G., Kross, E., Krpan, K. M., Askren, M. K., Burson, A., Deldin, P.J., Kaplan, S., Sherdell, L., Gotlib, I. H., & Jonides, J. (2012). Interacting with nature improves cognition and affect for individuals with depression. *Journal of Affective Disorders, 140*(3), 300–305.

The Effect of Training on Expertise

Ericsson, K. A., Krampe, R. T., & Tesch-Römer, C. (1993). The role of deliberate practice in the acquisition of expert performance. *Psychological Review, 100*(3), 363–406.

Newport, C. (2016). *Deep work: Rules for focused success in a distracted world.* London, England: Piatkus.

The Effect of Brain-Training Programs

Owen, A. M., Hampshire, A., Grahn, J. A., Stenton, R., Dajani, S., Burns, A. S., Howard, R. J., & Ballard, C. G. (2010). Putting brain training to the test. *Nature, 465*(7299), 775–778.

The Benefits of Meditation

Slagter, H. A., Davidson, R. J., & Lutz, A. (2011). Mental training as a tool in the neuroscientific study of brain and cognitive plasticity. *Frontiers in Human Neuroscience, 5.*

MacLean, K. A., Ferrer, E., Aichele, S. R., Bridwell, D. A., Zanesco, A. P., Jacobs, T. L., King, B. G., Rosenberg, E. L., Sahdra, B. K., Shaver, P. R., Wallace, B. A., Mangun, G. R., & Saron, C. D. (2010). Intensive meditation training improves perceptual discrimination and sustained attention. *Psychological Science*, 21(6), 829–839.

Mrazek, M. D., Franklin, M. S., Phillips, D. T., Baird, B., Schooler, J. W. (2013). Mindfulness training improves working memory capacity and GRE performance while reducing mind wandering. *Psychological Science*, 24(5), 776–781.

Jha, A. P., Krompinger, J., Baime, M. J. (2007). Mindfulness training modifies subsystems of attention. *Cognitive, Affective, & Behavioral Neuroscience*, 7(2), 109–119.

The Benefits of Going Offline

Perlow, L. A., Porter, J. L. (2009, October). Making time off predictable—and required. *Harvard Business Review*, 102–109.

The Right to Be Offline

NOS op3 (2017, January 1). Mailt je baas in de avond? In Frankrijk hoef je niet meer te reageren [Does your boss email you at night? In France, you no longer have to respond]. *NOS*. Retrieved from http://nos.nl/op3/artikel/2150987-mailt-je-baas-in-de-avond-in-frankrijk-hoef-je-niet-meer-te-reageren.html.

The Positive Effects of Exercise

Hillman, C. H,, Pontifex, M. B., Castelli, D. M., Khan, N. A., Raine, L. B., Scudder, M. R., Drollette, E. S., Moore, R. D., Wu, C.-T., & Kamijo, K. (2014). Effects of the FITKids randomized controlled trial on executive control and brain function. *Pediatrics*, 134(4), 1063–1071.

Ratey, J. J., & Loehr, J. E. (2011). The positive impact of physical activity on cognition during adulthood: A review of underlying mechanisms, evidence and recommendations. *Reviews in the Neurosciences, 22*(2), 171–185.

The Effect of Brain Stimulation on Attention and Concentration

Iuculano, T., & Kadosh, R. C. (2013). The mental cost of cognitive enhancement. *Journal of Neuroscience, 33*(10), 4482–4486.

The Effect of Medication on Attention and Concentration

Advokat, C. (2010). What are the cognitive effects of stimulant medications? Emphasis on adults with attention-deficit/hyperactivity disorder (ADHD). *Neuroscience & Biobehavioral Reviews, 34*(8), 1256–1266.

Chapter 5: The Importance of Concentration in Traffic

Koen van Tongeren Accident

Veroorzaker ongeluk: "Door mij leeft een kind van twee niet meer" [Cause of accident: "Because of me, a two-year-old child no longer lives"] (2015, April 2). *RTL Nieuws.* Retrieved from https://www.rtlnieuws.nl/nieuws/veroorzaker-ongeluk-door-mij-leeft-een-kind-van-twee-niet-meer

Tommy-Boy Accident

Van Weezel, T. G. (2016, September 1). "Tommy-Boy werd aangereden in een voor hem geweldige zomer" ["Tommy-Boy was hit during a great summer for him"]. *De Volkskrant.* Retrieved from https://www.volkskrant.nl/media/-tommy-boy-werd-aangereden-in-een-voor-hem-geweldige-zomer~a4368683/.

The Relationship between Accidents and Smartphone Use

Redelmeier, D. A., & Tibshirani, R. J. (1997). Association between cellular-telephone calls and motor vehicle collisions. *New England Journal of Medicine, 336,* 453–458.

Strayer, D. L., Drews, F. A., & Crouch, D. J. (2006). A comparison of the cell phone driver and the drunk driver. *Human Factors, 48*(2), 381–391.

The Eye Movements of Nico Hülkenberg

Schrader, S. (2016, July 3). This is how a formula one driver sees the track. *Jalopnik.* Retrieved from https://blackflag.jalopnik.com/ this-is-how-a-formula-one-driver-sees-the-track-1782965187

F1 driver eye tracking: Nico Hulkenberg tests out reactions (2016, August 16). *Sky Sports* Retrieved from http://www.skysports.com/ f1/news/24227/10328011/f1-driver-eye-tracking-nico-hulkenberg -tests-out-reactions

Reynolds, M. (2016, July 4). See an F1 race through the eyes of the driver and witness his near superhuman reactions. *Wired.* Retrieved from http://www.wired.co.uk/article/see-view-of-f1-driver-with-vision -tracking-technology

Different Types of Conversations While Driving

Drews, F. A., Pasupathi, M., & Strayer, D. L. (2008). Passenger and cell phone conversations in simulated driving. *Journal of Experimental Psychology: Applied, 14*(4), 392–400.

Strayer, D. L., & Drews, F. A. (2007). Cell-phone-induced driver distraction. *Current Directions in Psychological Science, 16*(3), 128–131.

Rueda-Domingo, T., Lardelli-Claret, P., de Dios Luna-del-Castillo, J., Jiménez-Moleón, J. J., García-Martín, M., & Bueno-Cavanillas, A. (2004). The influence of passengers on the risk of the driver causing a car collision in Spain: Analysis of collisions from 1990 to 1999. *Accident Analysis and Prevention, 36*(3), 481–489.

LED Lighting in Bodegraven

Pilotproject met LED-lichtlijnen bij oversteekplaats [Pilot project with LED light lines at crosswalks] (2017, February 9). *Rebonieuws.* Retrieved from https://www.rebonieuws.nl/uncategorized/pilotproject-led-lichtlijnen-oversteekplaats/.

Walking and Talking on a Smartphone in Japan

The Buzz (language column) (2014, March 8). Aruki-sumaho ("smartphone walking"). *Japan Times.* Retrieved from https://www.japantimes.co.jp/life/2014/03/08/language/aruki-sumaho/#.XShneOgzY2w.

Cathy Cruz Marrero

Daily Mail Reporter (2012, March 15). In deep water: Woman who fell into fountain while texting admits to spending thousands of dollars on stolen credit card. *Daily Mail.* Retrieved from http://www.dailymail.co.uk/news/article-2115438/Fountain-woman-Cathy-Cruz-Marrero-sentenced-months-house-arrest-shopping-spree-stolen-credit-card.html.

Masterson, T., & Stamm, D. (2011, January 21). Security guard who put fountain fall online gets fired. *NBC Philadelphia.* Retrieved from http://www.nbcphiladelphia.com/news/local/Foutain-Texter-Security-Firing-114398399.html.

Pedestrian Behavior in New York

Basch, C. H., Ethan, D., Rajan, S., & Basch, C. E. (2014). Technology-related distracted walking behaviours in Manhattan's most dangerous intersections. *Injury Prevention, 20*(5), 343–346.

Walking Behavior of Distracted Children

Stavrinos, D., Byington, K. W., & Schwebel, D. C. 2009. Effect of cell phone distraction on pediatric pedestrian injury risk. *Pediatrics, 123*(2), 179–185.

Chaddock, L., Neider, M. B., Lutz, A., Hillman, C. H., & Kramer, A. F. (2012). Role of childhood aerobic fitness in successful street crossing. *Medicine & Science in Sports & Exercise, 44*(4), 749–753.

Headphone Usage and Accidents

Lichenstein, R., Smith, D. C., Ambrose, J. L., & Moody, L. A. (2012). Headphone use and pedestrian injury and death in the United States: 2004–2011. *Injury Prevention, 18*(5), 287–290.

Multisensory Integration

Van der Stoep, N., Van der Stigchel, S., Nijboer, T. C. W., & Van der Smagt, M. J. (2016). Audiovisual integration in near and far space: Effects of changes in distance and stimulus effectiveness. *Experimental Brain Research, 234*, 1175–1188.

Van der Stoep, N., Nijboer, T. C. W., & Van der Stigchel, S. (2014). Exogenous orienting of crossmodal attention in 3D space: Support for a depth-aware crossmodal attentional system. *Psychonomic Bulletin & Review, 21*(3), 708–714.

Van der Stoep, N., Van der Stigchel, S., & Nijboer, T. C. W. (2015). Exogenous spatial attention decreases audiovisual integration. *Attention, Perception & Psychophysics, 77*(2), 464–482.

Safe Lock Study Fact-Check

Veldhuizen, R. (2017, June 26). Ook smartphonegeluidjes leiden fietsers gevaarlijk af—klopt dit wel? [Smartphone sounds dangerously distract cyclists—is this true?] *De Volkskrant*. Retrieved from https://www.volkskrant.nl/wetenschap/ook-smartphonegeluidjes -leiden-fietsers-gevaarlijk-af-klopt-dit-wel~a4502814/

De Volkskrant (2017, June 30). Klopt dit wel: appen tijdens het fietsen [Is this correct: Using smartphone while cycling]. YouTube video, 3:49. Retrieved from https://www.youtube.com/watch?v =lnAe2q1AIfQ&t=90s

Overconfidence in Traffic

Smit, P. H. (2017, August 2). Veilig Verkeer Nederland: gebruik smartphone in auto even ernstig als rijden onder invloed [Safe traffic in the Netherlands: Using a smartphone in the car as serious as driving under the influence]. *De Volkskrant.* Retrieved from https://www.volkskrant.nl/economie/veilig-verkeer-nederland -gebruik-smartphone-in-auto-even-ernstig-als-rijden-onder -invloed~a4509268/.

Chapter 6: The Future

The Fidget Spinner Craze

Brustein, J. (2017, May 11). How the fidget spinner origin story spun out of control. *Bloomberg.* Retrieved from https://www .bloomberg.com/news/articles/2017-05-11/how-the-fidget-spinner -origin-story-spun-out-of-control.

Singh, A. (2017, May 24). Fidget spinners: What is the new craze banned in schools across the nation? *Telegraph.* Retrieved from https://www.telegraph.co.uk/news/0/what-are-fidget-spinners -new-classroom-craze-banned-across-nation/.

de Vrieze, J. (2017, June 1). Helpt populaire fidget spinner tegen ADHD? [Does the popular fidget spinner help against ADHD?] *Elsevier Weekblad.* Retrieved from http://www.elsevierweekblad .nl/kennis/achtergrond/2017/06/helpt-populaire-fidget-spinner -tegen-adhd-92738w/.

Schneider, R. (2017, May 2). Fidget spinner manufacturers are marketing their toys as a treatment for ADHD, autism, and anxiety.

VICE. Retrieved from https://motherboard.vice.com/en_us/article/ 53nm5d/lets-investigate-the-nonsense-claim-that-fidget-spinners -can-treat-adhd-autism-and-anxiety.

Fidgeting and ADHD

Sarver, D. E., Rapport, M. D., Kofler, M. J., Raiker, J. S., & Friedman, L. M. (2015). Hyperactivity in attention-deficit/hyperactivity disorder (ADHD): Impairing deficit or compensatory behavior? *Journal of Abnormal Child Psychology, 43*(7), 1219–1232.

Interview about High Sensitivity with Fleur van Groningen

"Misschien is voelen weer toegelaten" ["Maybe feeling is allowed again"]. (2017, September 16). *De Standaard*. Retrieved from http://m.standaard.be/cnt/dmf20170915_03075747.

Elaine Aron and HSP

Aron, E. (2004, November 28). Is sensitivity the same as being gifted? *The Highly Sensitive Person* (blog). Retrieved from http:// www.hsperson.com/pages/3Nov04.htm.

The History of Jean Ayres

Heilbroner, P. L. (2015, November 9). Why "sensory integration disorder" is a dubious diagnosis. *Quackwatch*. Retrieved from https://www.quackwatch.org/01QuackeryRelatedTopics/sid.html.

Brain Activity and High Scores on the HSP Questionnaire

Acevedo, B. P., Aron, E. N., Aron, A., Sangster, M. D., Collins, N., & Brown, L. L. (2014). The highly sensitive brain: An fMRI study of sensory processing sensitivity and response to others' emotions. *Brain and Behavior, 4*(4), 580–594.

Owen, J. P., Marco, E. J., Desai, S., Fourie, E., Harris, J., Hill, S. S., Arnett, A. B., & Mukherjee, P. (2013). Abnormal white matter microstructure in children with sensory processing disorders. *NeuroImage: Clinical, 2*, 844–853.

The Official Statement of the American Academy of Pediatrics

Zimmer, M., & Desch, L. (2012). Sensory integration therapies for children with developmental and behavioral disorders. *Pediatrics, 129*(6), 1186–1189.

Multitasking and ADHD

Ewen, J. B., Moher, J. S., Lakshmanan, B. M., Ryan, M., Xavier, P., Crone, N. E., Denckla, M. B., Egeth, H., & Mahone, E.M. (2012). Multiple task interference is greater in children with ADHD. *Developmental Neuropsychology, 37*(2), 119–133.

ADHD and the Smartphone

van Noort, W. (2016, November 14). Zo schadelijk is afleiding door je smartphone [That's how harmful distraction is with your smartphone]. *NRC*. Retrieved from https://www.nrc.nl/nieuws/2016/11/14/allemaal-adhd-door-de-smartphone-4854932-a1531805.

Peak Attention and the TV

Wu, T. (2016). *The attention merchants: From the daily newspaper to social media, how our time and attention is harvested and sold.* New York, NY: Alfred A. Knopf.

The Relationship between Television Viewing and Reading and Attention Problems

Christakis, D. A., Zimmerman, F. J., DiGiuseppe, D. L., & McCarty, C. A. (2004). Early television exposure and subsequent attentional problems in children. *Pediatrics, 113*(4), 708–713.

Ennemoser, M., & Schneider, W. (2007). Relations of television viewing and reading: Findings from a 4-year longitudinal study. *Journal of Educational Psychology*, *99*(2), 349–368.

Hope for the Future

The Conscience-Stricken Executives of Silicon Valley
Verrycken, R. (2013, December 30). De spijtoptanten van Silicon Valley [The regrets of Silicon Valley]. *De Tijd*. Retrieved from https://www.tijd.be/tech-media/technologie/de-spijtoptanten -van-silicon-valley/9967875.html

Interview with Alan Lightman
van Noort, W. (2018, July 27). "Je moet véél meer lummelen en niksen": Alan Lightman, hoogleraar menswetenschappen ["You have to fiddle a lot more and do nothing": Alan Lightman, professor of human sciences]. *NRC*. Retrieved from https://www.nrc .nl/nieuws/2018/07/27/je-moet-veel-meer-lummelen-en-niksen -a1611414.

Index

Printed by Printforce, United Kingdom